Practical Batch Process Managem

Other titles in the series

Practical Cleanrooms: Technologies and Facilities (David Conway)

Practical Data Acquisition for Instrumentation and Control Systems (John Park, Steve Mackay)

Practical Data Communications for Instrumentation and Control (Steve Mackay, Edwin Wright, John Park)

Practical Digital Signal Processing for Engineers and Technicians (Edmund Lai)

Practical Electrical Network Automation and Communication Systems (Cobus Strauss)

Practical Embedded Controllers (John Park)

Practical Fiber Optics (David Bailey, Edwin Wright)

Practical Industrial Data Networks: Design, Installation and Troubleshooting (Steve Mackay, Edwin Wright, John Park, Deon Reynders)

Practical Industrial Safety, Risk Assessment and Shutdown Systems for Instrumentation and Control (Dave Macdonald)

Practical Modern SCADA Protocols: DNP3, 60870.5 and Related Systems (Gordon Clarke, Deon Reynders)

Practical Radio Engineering and Telemetry for Industry (David Bailey)

Practical SCADA for Industry (David Bailey, Edwin Wright)

Practical TCP/IP and Ethernet Networking (Deon Reynders, Edwin Wright)

Practical Variable Speed Drives and Power Electronics (Malcolm Barnes)

Practical Centrifugal Pumps (Paresh Girdhar and Octo Moniz)

Practical Electrical Equipment and Installations in Hazardous Areas (Geoffrey Bottrill and G. Vijayaraghavan)

Practical E-Manufacturing and Supply Chain Management (Gerhard Greef and Ranjan Ghoshal)

Practical Grounding, Bonding, Shielding and Surge Protection (G. Vijayaraghavan, Mark Brown and Malcolm Barnes)

Practical Hazops, Trips and Alarms (David Macdonald)

Practical Industrial Data Communications: Best Practice Techniques (Deon Reynders, Steve Mackay and Edwin Wright)

Practical Machinery Safety (David Macdonald)

Practical Machinery Vibration Analysis and Predictive Maintenance (Cornelius Scheffer and Paresh Girdhar)

Practical Power Distribution for Industry (Jan de Kock and Cobus Strauss)

Practical Process Control for Engineers and Technicians (Wolfgang Altmann)

Practical Power Systems Protection (Les Hewitson, Mark Brown and Ben. Ramesh)

Practical Telecommunications and Wireless Communications (Edwin Wright and Deon Reynders)

Practical Troubleshooting Electrical Equipment (Mark Brown, Jawahar Rawtani and Dinesh Patil)

Practical Hydraulics (Ravi Doddannavar, Andries Barnard)

Practical Batch Process Management

Mike Barker BSc (ElecEng), Senior Electrical Consultant, Johannesburg, South Africa

Jawahar Rawtani MSc (Tech), MBA, Senior Electrical Engineer, Nashik, India

Series editor: Steve Mackay FIE (Aust), CPEng, BSc (ElecEng), BSc (Hons), MBA, Gov. Cert. Comp., Technical Director – IDC Technologies

AMSTERDAM • BOSTON • HEIDELBERG • LONDON
NEW YORK • OXFORD • PARIS • SAN DIEGO
SAN FRANCISCO • SINGAPORE • SYDNEY • TOKYO

ELSEVIER

Newnes is an imprint of Elsevier

Newnes

Newnes
An imprint of Elsevier
Linacre House, Jordan Hill, Oxford OX2 8DP
30 Corporate Drive, Burlington, MA 01803

First published 2005
Copyright © 2005, IDC Technologies. All rights reserved

British Library Cataloguing in Publication Data
Barker, Mike
  Practical batch process management
  1. Process control   2. Process control—problems, exercises, etc.
  3. Electronic data processing—Batch processing.
  4. Electronic data processing—Batch processing—problems, exercises, etc.
  I. Title   II. Rawtani, Jawahar
  670. 4'27

Library of Congress Cataloguing in Publication Data
A catalogue record for this book is available from the Library of Congress

ISBN 0 7506 6277 8

For information on all Newnes publications
visit our website at www/newnespress/com

Typeset and edited by Integra Software Services Pvt. Ltd, Pondicherry, India
www.integra-india.com

Printed and bound in the United Kingdom

Transferred to Digital Print 2011

Contents

Preface

Historically, batch control systems were designed in individual ways to match the basic arrangement of plant equipment. They lacked the ability to convert to new products without having to modify the control systems. These schemes did not lend themselves to recipe-based operations or to integration with manufacturing management systems. This book concentrates on getting the building bricks right and arranging the structures into flexible schemes suitable for automated batch management e.g being able to work in response to new recipes that use the same plant equipment in different combinations. The material in this book aligns with current practices in the automation of batch processes, including the drive for integration with MES and ERP products from major IT product companies. References and examples will be drawn from Distributed Control Systems/Programmable Logic Controller batch control products in the market place.

This book shows you how to:

- Structure the activities of batch control into easy-to-understand tasks
- Choose, design and manage an automated batch management control system
- Save your business time and money by choosing and designing the correct, and therefore efficient batch process control system.

This book is suitable for:

- Engineers and technicians in process or control/instrument fields who are involved in batch process control projects
- Production supervisors or managers interested in developing improved batch management
- Techniques through the use of automation systems
- System integrators seeking to provide a design service to clients
- Those in businesses that have automated batch manufacturing as a part of their production activity.

It is hoped that the book will enable you to:

- Acquire a basic knowledge of standards and techniques in the automation of batch processes
- Design a simple batch manufacturing control system for new or upgrade projects using principles supported by the ISA S88 standard
- Obtain guidance in the integration of batch control systems with manufacturing information systems
- Design and specify simple instrumentation and batch controls in modules leading to complete unit operations
- Carry out the simple design of batch control operations including the sequencing and interlocking functions
- Develop batch operations into complete recipe-based production systems

- Be aware of and be able to evaluate the choices in the range of batch control system products
- Avoid the pitfalls of not having the batch control system package match your requirements.

You will need a basic understanding of instrumentation and process control systems for the book to be of greatest benefit. No previous knowledge of batch management is required but a modicum of practical experience with batch systems would be helpful to place the material into context.

1

Introduction

1.1 Introduction

The manufacturing industry has drawn its efficiency from large-scale continuous processes over a long period. Initially, the manufacturing facility for a new product used to be either a batch process or a laboratory process on a larger scale. But as the economy of scale was key to success in business, chemical engineering and process industries focused all attention on designing and developing continuous processes. Continuous processes are dominant in manufacturing of bulk chemicals. However, for manufacturing fine and specialty chemicals, with the increased emphasis on and customer requirements of high quality, equal focus has been on batch processing. Today, almost half of the processes in industry are batch processes.

Following are the economic and technical factors that make batch processes favorable over continuous processes:

- Batch processes often consist of simple processing units like mixers and stirrers.
- A batch processing unit may be multi-purpose – it may be used for several processing phases of the batch and could support multi-product manufacturing within the facility.
- Batch manufacturing plants are comparatively more robust than a continuous plant.
- Batch process manufacturing facility is easier to scale up depending on market demand and requirements.

1.1.1 Classification of processes

Industrial processes are classified depending on the output of the process as:

- Continuous process
- Discrete process
- Batch process.

Continuous process

In a continuous process, there is continuous flow of material or product. Processing the materials in different equipments produces the products. Each equipment operates in a single steady state and performs specific processing function.

Some examples of continuous processes are generation of electricity, Cement/Clinker production, paper mill and so on.

Discrete process

In a discrete process, the output of the process appears one-by-one or in discrete quantities. The products are produced in lots based on common raw materials and production history. In discrete process, a specified quantity of products moves as a unit or group of parts between workstations.

Some examples of discrete processes are assembly of watches, production of cars, assembly of television set, etc.

Batch process

In a batch process, the output of the process appears in quantities of materials or lots. A batch process has a beginning and an end. Batch processes are neither continuous nor discrete, but have the characteristics of both. Batch process is usually performed over and over. The product of a batch process is called a batch. Batch processes define a sub-class of sequential processes. Batch processes generate a product but the sequential processes need not necessarily generate a product.

Some examples of batch processes are beverage processing, biotech products manufacturing, dairy processing, food processing, pharmaceutical formulations and soap manufacturing.

1.2 Identification of batch processes

A batch is defined as:

- The material that is produced by a single execution of a batch process, or
- An entity that represents the production of a material at any point of time in the process.

Here, it is important to note that the term 'batch' means both the material produced by and during the process, and also an entity that represents the production of that material. The term 'batch' is used as an abstract contraction for the words – the production of a batch.

We have seen classification of processes based on their outputs. Now we will formalize the definition of a batch process:

- A process is considered to be a batch process if, due to physical structuring of the process equipment or other reasons, the process consists of a sequence of one or more steps that must be performed in a defined order. And, on completion of the sequence of the steps a finite quantity of the finished product is produced. The sequence is repeated to produce another batch of the product.
- A batch process is a process that leads to the production of finite quantities of material by subjecting quantities of input raw materials to an ordered set of processing activities over a finite period of time using one or more equipments.

1.2.1 Batch processes

Today many products are produced using continuous processes that were originally produced using batch processes. The main reason behind the shift from batch process to continuous process was that the batch processes were labor-intensive and required skilled and experienced operators to produce batch products with consistency in quality. However, due to increasing demand for flexible and customer-driven production, batch processes find equal importance in manufacturing industries. Batch processes are economical for small-scale production as it requires few number of process equipment and intermediate storage is inexpensive. Batch processes are suitable for manufacturing of large number of products or special products due to flexibility in manufacturing process equipment.

Nature of batch processes

Batch processes have the following characteristics:

- Batch processes deal with discrete quantities of raw materials or products.
- Batch processes allow the tracking of these discrete quantities of materials or products.
- Batch processes allow more than one type of product to be processed simultaneously, as long as the products are separated by the equipment layout.
- Batch processes entail movement of discrete product from one processing area to the other.
- Batch processes have recipes (or processing instructions) associated with each load of raw material to be processed into product.
- Batch processes have more complex logic associated with processing than is found in continuous processes.
- Batch processes often include normal steps that can fail, and thus also include special steps to be taken in the event of a failure.

Nature of steps involved in a batch process is:

- Each step can be simple or complex in nature, consisting of one or more operations.
- Generally, once a step is started it must be completed to be successful.
- It is not uncommon to require some operator approval before leaving one step and starting the next.
- There is frequent provision for non-normal exits to be taken because of operator intervention, equipment failure or the detection of hazardous conditions.
- Depending on the recipe for the product being processed, a step may be bypassed for some products.
- The processing operations for each step are generally under recipe control, but may be modified by operator override action.

A typical batch process step is shown in Figure 1.1.

1.2.2 Classification of batch processes

Batch processes can be classified on the basis of two criteria:

1. The quantity of output produced
2. The structure of batch process plant.

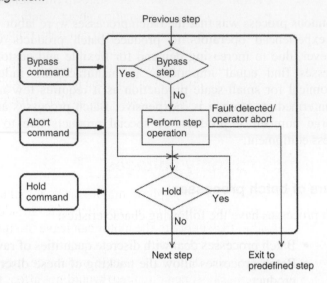

Figure 1.1
Typical batch process step

The number of products produced

- *Single-product batch process*:   Same amount of raw materials are used and same operations are performed on each batch to produce same product.
- *Multi-grade batch process*:   Same operations are performed on each batch with different amount of raw materials and/or under different processing conditions to produce similar but not identical products.
- *Multi-product batch process*:   Different methods of operations or control are performed on different amount of raw materials under different processing conditions to produce different products.

The structure of batch process plant

- *Single-path batch process*:   In a process with single-path structure, the batch passes sequentially in a predefined path from one unit to another, as shown in the Figure 1.2(a).
- *Multi-path batch process*:   In a process with multi-path structure, there may be several batches active at a time and the equipment may be of different physical characteristics. For example, as shown in Figure 1.2(b), a reaction operation may be handled by one unit in one path of the process and by two units in the other path of the process.
- *Network batch process*:   In a process with network structure, the sequence of the units may be pre-assigned, or determined prior to execution of the batch or during the execution of the batch. In network structure process, an appropriate path is determined at the time of execution depending on constraints like recipe requirements and equipment capabilities. A typical network process is shown in the Figure 1.2(c). Control of network process is complex due to the need for allocation of equipment and the arbitration of requests for the equipment.

Whatever be the structure of a batch process plant, the nature of the batch process involved and the types of equipment used or the operations involved, the objective is to manufacture batches of product. As discussed above, there are many variations – single plant, single- and multi-stage processes, etc. Whatever the plant configuration, consideration of any single batch reveals that a series of operations in a dedicated vessel or train of vessels according to some recipe always processes it. Hence, the starting point for developing a methodology for batch process control is to identify distinct products or generically similar products. After identification of products, for each product the route that a batch follows through the plant is to be established, and the operations carried out on that batch considered, i.e. which operations, where and in what order. Analysis of the structure of a batch plant gives a good insight into the nature of both the operations and the control requirements. The structure relates to the number and type of process streams, and to the trains of equipment used. A plant may be considered as a process cell consisting of a cluster of units and equipment modules within which the process stages are carried out. These units and equipment modules are organized either in series and/or in parallel. A stream is defined to be the order of units and equipment modules used in the production of a specific batch. Or in other words, a stream is the route taken through the plant by a batch. A train consists of the units and equipment modules used to realize a particular stream.

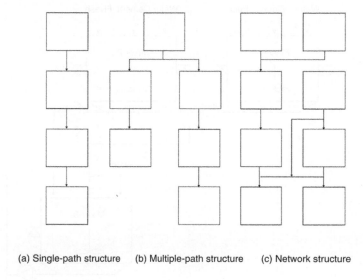

    (a) Single-path structure   (b) Multiple-path structure   (c) Network structure

Figure 1.2
Classification of batch processes based on structure of the plant

The complexity of a batch process plant based on the process structure and number of products is shown in the form of a matrix in Figure 1.3. The matrix indicates the degree of complexity involved in automation of various combinations. More complex a process, more it requires allocation, arbitration and batch management solutions. More the number of products being manufactured by a process, more is the need for recipe management and batch management solutions. A single-product, single-path batch plant is simple, while a multi-product, network structure is the most complex combination.

Example 1.1:

A simple example of batch process can be taken from fiber–cement products industry, wherein fiber, cement, pulp and additives are used in different proportions to prepare raw material slurry batches for making corrugated roofing and flat boards. A typical P&ID for the batch process is shown in Figure 1.4(a). The raw material preparation batching system can also be used to prepare slurry for fiber–cement high-pressure pipes.

Network	High	High	High
Multiple path	Low	Medium	High
Single path	Low	Medium	High
	Single	Few	Many

Process structure →

Complexity →

Number of products (recipes) →

Figure 1.3
Complexity of batch process plants

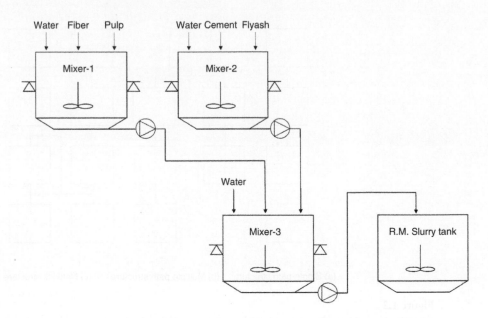

Figure 1.4(a)
Example of batch process – preparation of raw materials slurry in fiber–cement products plant

Example 1.2:

Another example of a batch process as shown in Figure 1.4(b), is a simplified plant for batch production of industrial ethanol. Ethanol is produced as a result of the biodegradation of glucose by yeast in a fed-batch mode. The plant consists of three fermentors, three product tanks, a feed tank, a sterilizer, pumps and valves.

Example 1.3:

An example of network structure in a brewery plant for fermentation batch process is shown in Figure 1.4(c). The plant consists of several high-volume tanks that are interconnected by a complex network of pipes, valves and pumps. One of the processes taking place is fermentation of the worth. The brewing tanks deliver the worth to the fermentation tanks. After fermentation, the 'green beer' is pumped to aging tanks. Filling and emptying of a fermentation tank takes several hours and the fermentation process itself takes several days. The tanks are divided into clusters, as shown in Figure 1.4(c). There are several shared resources in the system; many tanks share the same pipes, pumps, cleaning equipment and processing equipment. Figure 1.4(c) illustrates groups of three-fermentation tanks that share the same pipe. This means that only one of these three tanks can perform an action using the pipe at the same time. Some actions are filling, emptying and cleaning.

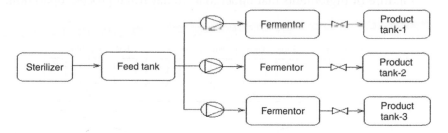

Figure 1.4(b)
Example of batch process – production of industrial ethanol

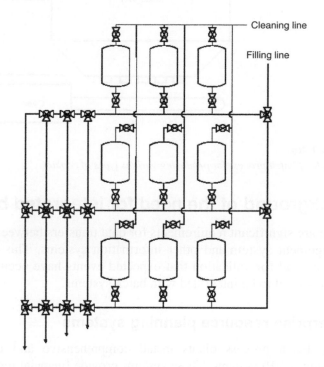

Figure 1.4(c)
Example of network structure batch process – fermentation process in a brewery plant

Example 1.4:

An example of batch process with time-based operations that produce various types of cookies is shown in Figure 1.4(d). Let us consider that the chocolate-chip cookies are made in the first production run. First, the oven is turned on to the desired temperature. Next, the required ingredients in proper quantities are dispensed into the sealed mixing chamber. A large blender then begins to mix the contents. After a few minutes, vanilla is added, and the mixing process continues. After a prescribed period of time, when the dough is in proper consistency, the blender stops turning, and the compressor turns on to force air into the mixing chamber. When the air pressure reaches a certain point, the conveyor belt turns on. The pressurized air forces the dough through outlet jets onto the belt. The dough balls become fully baked as they pass through the oven. The cookies cool as the belt carries them to the packaging machine. After the packaging step is completed, the mixing vat, blender and conveyor belt are washed before a batch of raisin–oatmeal cookies is made. Products from foods to petroleum to soap to medicines are made from a mixture of ingredients that undergo a similar batch process operation.

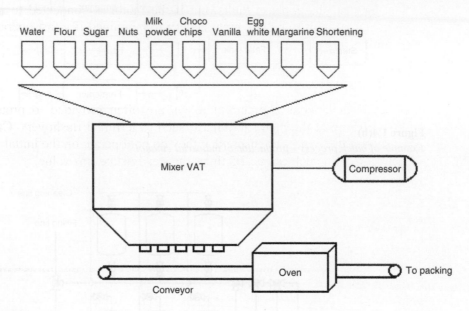

Figure 1.4(d)
Example of batch process for producing various types of cookies

1.3 Background of the need for integrated batch systems

There are significant requirements for data transfers between the batch process operations management system and other information systems. This data is used for calculations, tracking and for validation that expected events have occurred. Normally, the following systems need to be integrated with batch systems.

1.3.1 Enterprise resource planning systems

Many batch process plants install comprehensive and integrated enterprise resource planning (ERP) systems. These systems provide financial reporting, order entry, warehouse monitoring and shipping capabilities. They do not replace the operations management systems. The operations management system is required to provide the validated production data required for the ERP system. The ERP system will download customer order data and average raw material requirements. These are translated into specific scheduled grade productions and shipments by the operations system. The operations system also transfers the reconciled production data for long-term storage to the ERP system.

1.3.2 Maintenance systems

Integration of the maintenance system with the operations management system is important for tracking equipment utilization and to coordinate time information of equipment in-service and out-of-service.

1.3.3   Lab systems

For quality assurance and quality control of products, close integration of the operations management system with the laboratory system is required. Product samples and quality measurements are time-stamped and stored with other batch data.

1.3.4   Supply information system

Raw material quantities and qualities that have been recorded by the operations system are compared and balanced to those reported through the supply system to identify discrepancies.

1.3.5   Distribution information system

Shipment quantities and qualities are calculated by the system and are compared to those reported by distribution systems.

The batch process and its control have evolved over a long period of time from mechanical devices used to regulate the levels in liquid tanks, followed by pneumatic controls to the modern electronic controls based on microprocessor-based controllers and expert control systems. With the availability of sophisticated measurement sensors, advanced control systems and reliable communication technology, the demand for integrated control systems has increased. With the use of computers and software-based control algorithms, use of complex control algorithms apart from conventional PID control, ratio control, cascade control, etc. has increased in manufacturing industries including that for batch processes control. Nowadays, computer-based control systems are used for process control and optimization of processes. A computer has capabilities to monitor, compute and analyze multitude of variables and take multiple actions simultaneously.

Process automation helps in managing processes in all process conditions. Control systems help in coordination of activities of the entire process and respond to dynamic changes in the process. With the help of artificial intelligence-based expert systems, it is possible to optimize a process and manage it without manual interventions. Integrated batch automation systems provide process control closer to the ideal expectation of the operator with highest flexibility.

As shown in Figure 1.5, adoption of integrated batch management system as per the ISA S88 standard requires discipline but provides modular approach, reduces complexity and increases the flexibility of the process.

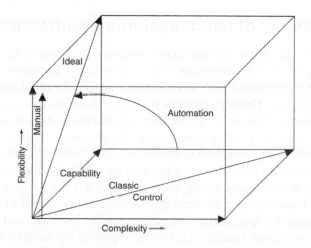

Figure 1.5
Process capabilities with integrated automation

To meet the increasing demand for integrated batch control systems, batch control models are combined with information models to cater to the needs of batch plant information. The main objective of an information model is to establish a proper framework for information management systems that support integrated batch plant operations.

In the 1980s, the computer-integrated-manufacturing (CIM) model represented use of information technology in manufacturing. The CIM model represented process control system devices to the business systems. But then organizations moved towards de-centralization and flat organization structures. Therefore, hardware-based CIM model became obsolete and software-based CIM model with three levels as shown in Figure 1.6 was evolved. The top-most level of the model pyramid was the strategic level constituting strategic planning and included business functions like sales, finance and manufacturing resource planning. The middle level constituted tactical level comprising of plant management functions like process, production, maintenance, resources and quality management. And the bottom level was operational level that covered the process control and the manufacturing control functions.

Figure 1.6
Computer-integrated-manufacturing (CIM) model levels

1.4   Overview of batch systems engineering

For a batch process (or any other process), the process life cycle is always longer than the life cycle of the automation system. Normally, a process outlives the automation due to fast developments and advancement in the field of automation and process control technologies. During a process life cycle there may be several generations of automation, as shown in Figure 1.7. Depending on process life cycle and process requirements, it is at times necessary to decommission the obsolete system and upgrade it with the latest generation of automation systems.

Automation upgrades have become increasingly important in the industrially developed world, where many aging control systems exist. Though functioning satisfactorily for years, many of these systems do not meet today's need of sophisticated control, historical recording for traceability, human–machine interface and supply chain synchronization. These outdated systems need to be replaced by modern batch control systems, on open platforms to achieve current regulatory requirements and current business performance needs.

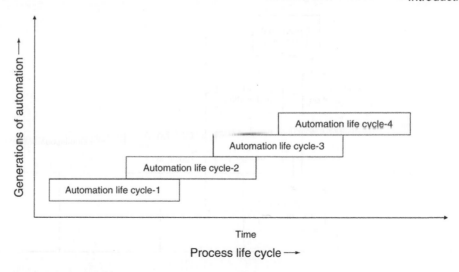

Figure 1.7
Automation life cycles during process life cycles

Improving operational efficiency is fueling the growth. Manufacturing efficiency and flexibility is key to meeting the needs of powerful mass merchandisers. The consumers demand more variety and best quality. To maximize business performance, a manufacturing company has to optimize its production plants along with its enterprise and supply chain domains. The need to optimize the supply chain is increasing the need for real-time plant and production information, which is fueling the growth of manufacturing automation and its integration to business system. Offering optimization at all levels is no longer sufficient for an enterprise to have islands of automation along with isolated supply chains. To maximize the potential of batch production, manufacturing plants and supply chains need to work together in an optimized fashion. In such environment, manufacturing must continue to maintain its central role, but enterprise-wide optimization of production facilities and supply chains are equally important for maintaining the bottom line. Manufacturing enterprises are optimizing not only their production plants but also their enterprise and supply optimization. Manufacturers used to be satisfied when control loops were properly tuned, recipes, phases ran in proper order, and products were produced to their required quality specifications. With today's global environment, increased competition and the need for custom products, enterprise and supply chain optimization are becoming increasingly important.

As shown in Figure 1.8, the strategic batch automation/system project engineering has six phases:

1. Definition of requirements
2. Design
3. Implementation
4. Testing and startup
5. Continuous evaluation and improvement
6. Decommissioning.

1.4.1   Definition of requirements

In the first stage, the batch automation/system objectives are clearly identified and defined. The requirements defined must be communicated to all the members of the team and must be understood by all.

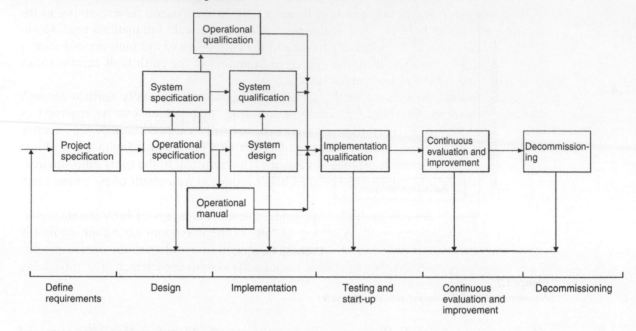

Figure 1.8
Phases of batch automation/system engineering

Following are the main objectives of defining the requirements:

- Document the scope of the batch process
- Develop the objectives for the batch process that are specific, measurable and achievable
- Develop the objectives for the batch automation system that are specific, measurable, achievable and aligned with the objectives of the batch process
- Define the parameters, against which the performance of the project will be evaluated
- Begin with a clear-cut target in mind
- Involve all the team members.

1.4.2   Design

During design, the operational specification is written without any mention of the platform or how the control programming is to be done. The operation of the process is defined based on the process capabilities and not based on the product needs. The ISA SP88 standard clearly distinguishes between the process and the product, i.e. between the equipment and the procedure. For existing process, P&IDs can be designed during this phase; otherwise design should be done with PFDs. During design phase, it is defined how the P&IDs should be organized and the measurement and output needs of the process. The system and the software to be used are also defined. During the design phase, after all the required inputs are available, safety and hazard analysis and operability studies can be also be conducted.

Following are the main objectives of design phase:

- Modularize process
- Define operational requirements based on which operational specification will be prepared
- Simplify and optimize process
- Define strategy for alarm management system

- Define process communication
- Define system requirements based on which the system specification will be made
- Selection of system based on system requirements.

1.4.3 Implementation

During implementation phase, the process design activities like preparation of layout, piping design and P&IDs for a new process are developed. Depending on the requirement, construction also takes place during this phase. The batch automation system is designed and implemented during this stage. The control system is configured based on the process requirements and programmed to perform the desired tasks. Implementation of control strategies is documented.

Following are the main objectives of the implementation phase:

- Implement the control strategies as defined during design phase
- Develop man–machine interface
- Create system qualification acceptance criteria
- Documentation of control strategies.

1.4.4 Testing and startup

The automation system is installed and tested as per the plan and procedure defined during the design and implementation stages. Simulation techniques are used for testing before startup. After successful testing, next step is to startup the process. If required, process validation is also done during this stage.

Following are the main objectives of this phase:

- Carry out simulation
- Test the control strategies as defined during the design and implementation stages
- Train the operators
- Deliver a batch control/management system that works.

1.4.5 Continuous evaluation and improvement

During this phase the system is operational. The performance of the system is monitored and evaluated continuously and improvements are carried out in the system after re-evaluation of previous phases.

The main objectives of this phase are:

- System assurance
- Continuous upgradation of system
- Backup of the system.

1.4.6 Decommissioning

The system is decommissioned and removed if the system becomes obsolete and is to be removed or any major changes are required in the process. Before decommissioning, all previous phases must be evaluated for the necessary changes. System documentation is updated.

The main objectives of the decommissioning phase are:

- Evaluate all previous phases for impact of changes to be made
- Evaluate impact on upstream and downstream processes
- Update system documentation.

1.5   Introduction to standards

Why was the need felt for batch management standard?

- No universal model existed
- Confidence that such a model could be created
- Help understand the big picture
- Tie in the details with the big picture
- Allow better communications

  – Among staff
  – With vendors

- Reduce risk WRT planning, costing and project management
- Basis for better documentation
- Allow for simulation.

1.5.1   The Instrument Society of America (ISA) S88 standard

The S88 standard was approved by the Instrument Society of America (ISA) in 1995 and later adopted as an IEC standard (IEC 61512-1). The S88 standard establishes a framework for the specification of requirements for batch process control, and for their subsequent translation into application software. The framework consists of a variety of definitions, models and structures.

Part 1 of the S88 standard on batch control defines a consistent set of terminology and models for defining and describing the control requirements for batch manufacturing processes/facilities. The SP88 Standard Committee had developed the ISA S88.01 standard that provides batch manufacturers methods and tools required to analyze the existing batch processes and add new products. Today, batch control systems based on ISA S88 standard models are readily available for use with software packages.

The objectives of S88 standards are:

- To provide a common, consistent model for design and operations of batch manufacturing processes/facilities and batch control systems
- To continuously improve controls and efficiencies in batch manufacturing processes
- To improve communications.

The S88 standard terminology and models help you to define the available equipment, recipes and the steps involved in batch manufacturing of products. There are four types of models given in S88 standard. These are:

1. *Physical model*:   Used to define the hierarchy of equipment used in the batch manufacturing process/facility.
2. *Control activity model*:   Used to define the relationships between various control activities required for batch processing.
3. *Procedural control model*:   Used to define the control that enables the equipment to perform a task.
4. *Process model*:   Used to define the results of performing procedural control on the equipment in the batch process.

Figure 1.9 illustrates the relationship between ISA S88 standard model and terminology. These models will be discussed in detail in the next chapters.

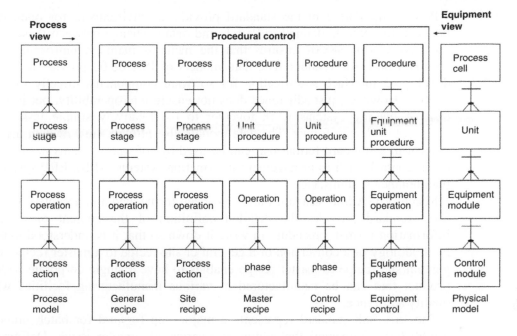

Figure 1.9
ISA S88 standard models and terminology

The S88.01 standard contains six sections:

1. Scope.
2. Normative references.
3. Definitions.
4. *Batch processes and equipment*: This section describes a physical model consisting of equipment objects – enterprise, site, area, process cell, unit, equipment module and control module, as shown in Figure 2.1.
5. *Batch control concepts*: This section deals with concept of basic, procedural and coordination control. This section describes procedural control model, which is made up of procedural entities like procedure, unit procedure, operation and phase. This section defines four types of recipes, namely general, site, master and control recipe. Production plans and schedules, resource allocation, production information, modes and states and exception handling are also included in this section.
6. *Batch control activities and functions*: This section describes the functions and activities necessary to make the pieces defined in Sections 4 and 5 to work together. This section is based on the control activity model.

ISA S88 standard Part 2 defines the relationships described in the control activity model of S88.01 standard and the data passed between those activities through data modeling. The scope of S88 Part 2 is:

- To provide a data model that represents the recipe, batch schedule, batch history and equipment objects
- To provide data exchange formats for recipe, schedule, history and equipment objects
- To define language guidelines to support data exchange of components of S88.01
- To define language guidelines for user representation of procedural elements.

Part 1 and Part 2 of the standard provide an environment for open data exchange between the objects defined in Part 1, and data exchange between these objects and the business world. S88.02 ensures that the methods maintain uniformity and provide a common base for vendors. The data models of ISA 88.02 provide a starting point for developing interface specification for software components directly involved in batch control. However, just adherence does not ensure interoperability, but it does reduce the task of making systems work together to the relatively simple task of deciding on technical exchange format. The data models also provide view of data involved in batch control in a standard format that can be used by software systems not directly related to batch control, such as management information systems (MIS), ERP systems, preventive maintenance systems, etc.

Recently ISA S88 standard Part 3 has been formulated and accepted. Part 3 of the ISA S88 standard provides the detailed definition of the content of a general recipe, information format, procedure to write it down so that it is understood universally and a methodology of a conversion of a general or site recipe to a master recipe. Conversion of general/site recipe to master recipe enables to have a single corporate-wide recipe for a product that can be manufactured at various manufacturing facilities with consistent quality assurance.

The ISA S88 standard represents a major step forward for batch process control; it establishes a framework that addresses a variety of complex issues. Defining terminology and relating it to meaningful models removes the ambiguity. The structures involved lead to the development of more comprehensive specifications, which consequently lead to better quality application software. The control system manufacturers and suppliers evaluate and review the control functions and structures of their systems to ensure conformance with the S88 standard requirements. Suppliers develop tools to enable users to articulate their requirements in terms of S88 objects and constructs, and for those requirements to be translated automatically into application software. Detailed decomposition of the various activities goes a long way to this end. Unlike most of the standards, which are retrospective in the sense that they attempt to provide a reference based on presently accepted good practices, the ISA S88 standard models and structures provide a flexible framework with much scope for context-based interpretation and application.

2

Identify and define physical models

2.1   Introduction

Physical models are used to describe the physical assets of an enterprise and the hierarchical relationships among the various assets involved in batch manufacturing. Usually, physical assets are organized in a hierarchical structure. Physical model provides a means to organize and define the equipment used to control the batch process. The physical model has seven hierarchical levels, starting from 'Enterprise' at the apex to 'Control modules' at the bottom of the hierarchy. A typical hierarchical structure of a physical model is shown in Figure 2.1. The S88 standard scope covers the bottom four levels of the physical model, as shown in Figure 2.1. The top three levels, namely, enterprise, site and area are not addressed by the ISA S88 standard. A brief description of these levels is as follows.

2.1.1   Enterprise

Enterprise defines the company that owns the plant or manufacturing facility.

2.1.2   Site

Site defines the location of the plant or manufacturing facility. The enterprise and the site levels provide a link to business systems and regulatory compliance requirements.

2.1.3   Area

Area defines a specific section of the site. The area contains one or more process cells.

In a physical model, the bottom four levels are specific to the types of equipment. Higher level consists of groups of lower levels in the hierarchy.

Hierarchy levels are defined by engineering activities. During the engineering activity, to simplify the operation of an equipment, all the equipments at the lower levels are grouped together to form a higher-level grouping and the equipment is treated as a large piece of equipment. The bottom four levels of the physical model are described below and illustrated in Figure 2.2.

Process cell

A process cell is the span of logical control of a set of process equipments within an area used to manufacture a batch product. The product produced may not be a final or finished product (it may be an intermediate product). Process cell helps in designing control actions and functions and also helps in production planning and scheduling on a process cell basis. Process cell is the domain for batch control system. A process cell must contain at least one unit, which is usually centered on a major equipment item, such as a mixing vessel or a reactor.

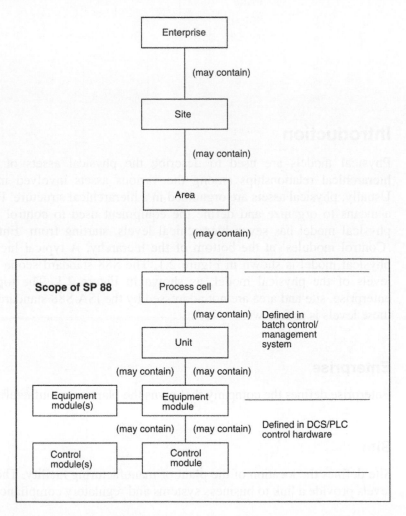

Figure 2.1
Hierarchical structure of physical model

Unit

A unit can carry out one or more major processing activities such as mixing, refining, reacting, blending, separation, etc. A unit is a collection of equipment modules and control modules. The equipment and control modules may be dedicated to the unit or may be temporarily acquired for carrying out the major processing activity for the unit.

A unit operates relatively independently of other units. It includes all logically related equipments necessary to perform the processing required. Although a unit frequently operates on, or contains, the complete batch of material at some point in the processing cycle, it may operate on only part of a batch, i.e. a partial batch. However, a unit may not operate on more than one specific batch at any point in time. Put another way, different parts of one unit cannot simultaneously contain parts of different batches. It follows that defining the boundary of a unit is a key design decision. Units normally, but not necessarily, consist of equipment modules, which may themselves contain other equipment modules.

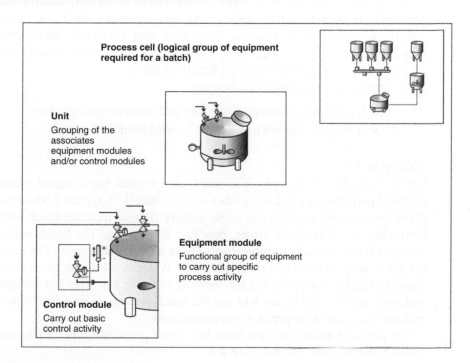

Figure 2.2
Illustration of physical model

A unit may have the following key attributes:

- Operates on full or a part of the batch
- Operates only one batch at a time
- Operates independently of other units and cannot acquire another unit
- Consists of flexible amount of equipment, equipment modules and control modules.

Equipment module

An equipment module can carry out a finite number of minor processing activities like analyzing and classifying, weighing and dozing, etc. An equipment module is a functional group of modules that carry out specific minor processing activities. It combines all necessary physical processing and control equipment required for performing these activities.
   An equipment module may have the following key attributes:

- Centered around an equipment, such as an agitator, heat exchanger, pre-heater cyclone, pre-calcinator, etc.
- Its scope is defined by the finite processing activity it is designed for
- Consists of equipment, other equipment modules and control modules

- Contains all equipment and control functions necessary to perform its process function
- May be part of a process cell, unit or another equipment module.

Control module

A control module is the lowest level in the hierarchy of a physical model. Although, formally not included as a level in the physical model, control modules consist of elements. It operates as a single entity and performs the basic control function. It is usually a collection of sensors, controllers and actuators. Basic control may include regulatory control or sequential control involving state-based control. It includes exception handling and monitoring functions also.

Key attributes of a control module are:

- It consists of actuators, sensors and other control modules
- It may be formed using other control modules.

Example 2.1:

Let us consider an example of a simple and typical batch manufacturing process of raw material preparation for making fiber–cement sheets. A typical fiber–cement manufacturing plant consists of three process cells, namely (i) raw material batch preparation, (ii) sheet formation and (iii) sheets sizing, handling and curing. The batch process of making raw material slurry to make roofing sheets or flat sheets is common. The raw materials required are fiber, pulp, cement and other additives. Water is used with these raw materials to make slurry in batches for producing the sheet on sheeting machine. For the sake of simplification and easy understanding, we will use the batch process of preparing raw material slurry for making sheets as an example in our discussions.

The physical model for raw materials batch preparation plant for manufacturing fiber–cement sheets is shown in Figure 2.3.

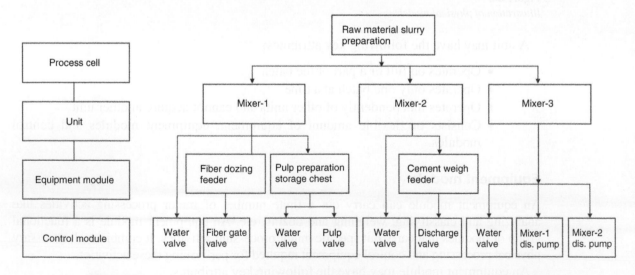

Figure 2.3
Physical model for raw materials batch preparation plant for manufacturing fiber–cement sheets

Example 2.2:

Another example of a physical model for polymerization train is shown in Figure 2.4.

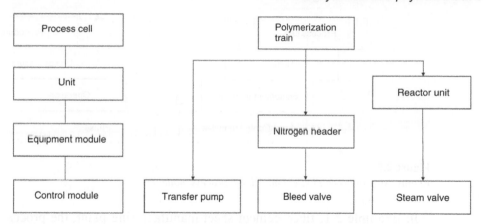

Figure 2.4
Physical model for a polymerization train

2.2 Define the physical model

Identifying and defining the physical model is a critical part of designing and implementing a batch management system. It is important to divide the batch process into modules. In the ISA SP88 standard, for control purposes it is assumed that the process cell for both the physical equipment as well as the related control activities, are clearly sub-divided into units, equipment modules and control modules. Sub-division of the process cell requires clear understanding of the functions and purpose of the process cell equipment. It also requires identification of equipment entities such as units, equipment modules and control modules that must work together to carry out the required processing functions by the process cell.

However, an effective sub-division of process cell into well-defined units, equipment modules or control modules is a complex activity. It depends on the specific requirements and the environment of the batch process involved. Improper or ineffective sub-division of process cell equipment entities can lead to a compromise on effectiveness of the modular approach to recipes as per the SP88 standard.

In the previous section, we have described the physical classification of the process into process cells, units, equipment modules and control modules. Such physical classification or partition should always be done from top to down. First the process should be identified with its boundaries. Then, identify the units within the process cell. Next step is to identify the equipment modules for each unit and the process cell. And finally control modules for each equipment module, units and the process cell are identified. This top to down approach for physical classification of the process and relationship between the physical model and process model is illustrated in Figure 2.5.

2.2.1 Identify and define process cells

The process cell, as defined in the previous section, consists of process entities – units, equipment modules and control modules. A process cell may include trains. 'Train' is a term used for collection of units with associated equipment modules and control modules that can be used to make a complete batch. Units in the process cell need not be physically connected. It is not necessary that the batch being executed uses all the equipments defined in the process.

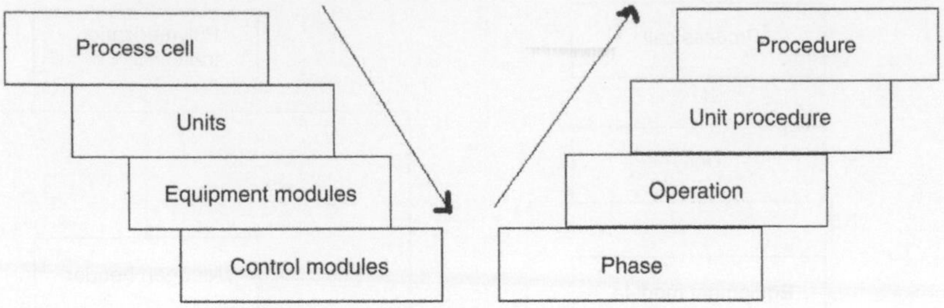

Figure 2.5
Relationship between physical model and process model

In Example 2.1, fiber–cement sheet manufacturing plant, the process of fiber–cement sheet manufacturing is sub-divided, defined and identified in three process cells, namely:

1. Raw materials slurry batch preparation
2. Sheet formation
3. Sheet sizing, handling and curing.

We will concentrate on raw materials slurry batch preparation cell for further sub-division of process cell into equipment entities in our subsequent discussions.

Following are some of the advantages of defining process cells:

- *Paths are predefined*:   At any point, if there are parallel units the choices are limited by the automatic or manual allocation.
- *Convenient grouping*:   Physical location may result in convenient groupings.
- *Choice limitation*:   For manufacturing a product, if a particular combination of units is recommended, though other units within the process cell can meet equipment criteria, the choices will be limited by the process cell.
- Limit on operational access.

2.2.2   Identify and define units

As defined in the previous section, a unit carries out one or more major processing activities. A unit is made up of equipment modules and control modules. Some of the equipment and control modules are designed to be permanent parts of the unit and some of the modules may be acquired depending on the requirement. A unit should be identified and defined such that it may operate independently. Typically, a unit contains or holds the batch for some processing. A unit, by definition, cannot execute two batches.

However, a unit can temporarily acquire another equipment module or control module to carry out specific activities.

In our Example 2.1 of a typical batch process, raw materials slurry batch preparation process cell, three units – Mixer-1, Mixer-2 and Mixer-3 have been identified and defined, as illustrated in Figure 2.6. The Mixer-1 unit carries out-processing on a batch of fiber and pulp with water. In Mixer-2 unit, cement slurry is prepared. And finally, a batch prepared by Mixer-1 unit and a batch prepared by Mixer-2 is processed together in Mixer-3 unit. Slurry prepared in Mixer-3 after stirring is a batch of raw materials slurry, which is supplied to a stuff chest with agitator for sheeting machine. The major processing activity carried out in the mixers is stirring.

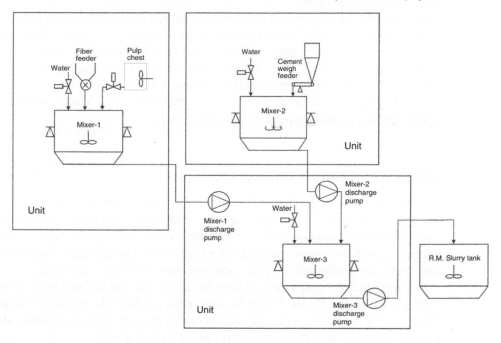

Figure 2.6
Example identification of units in a process cell

2.2.3   Identify and define equipment modules

An equipment module can carry out a finite number of minor processing. An equipment module consists of control modules and may also contain other equipment modules. Equipment modules can acquire the services of another equipment module or control module to carry out specific tasks. As an equipment module can be a shared resource, it can be designed and configured to operate on more than one batch at the same time. Some examples of shared equipment modules are vacuum system, Dust Collector Bag Filters, Hydraulic power pack, etc. An equipment module is capable of executing procedural control or phases. Equipment modules cannot act on a batch independently; units to perform a specific task use them.

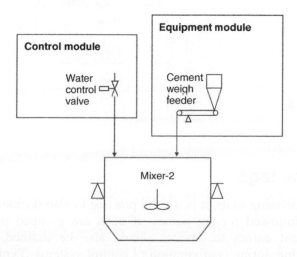

Figure 2.7
Example – identification of equipment module and control module

In Example 2.1, equipment modules are identified and defined as follows:

In Mixer-1 unit, there are two equipment modules – (i) fiber dozing feeder, (ii) pulp preparation and storage chest and one control module namely, water control valve.

As shown in Figure 2.7, the Mixer-2 unit consists of one equipment module, i.e. cement weigh feeder and one control module namely, water control valve.

Mixer-3 unit consists of three control modules – (i) Water control valve, Mixer-1 discharge pump and (iii) Mixer-2 discharge pump.

2.2.4   Identify and define control modules

A control module operates as a single entity and performs the basic control function. A control module is made up of sensors, controllers, actuators or other control modules. Control modules cannot execute procedural control; they are called upon by procedural control to perform specific action. As a control module can be a shared resource, it can be designed and configured to operate on one or more batches at a time.

In Example 2.1, equipment modules are identified and defined as follows:

As shown in the Figure 2.8, in Mixer-1 unit, for fiber dozing feeder equipment module, fiber hopper discharge gate valve is a control module. Water control valve and the pulp control valve are the other two control modules in the unit.

In Mixer-2 unit, for cement weigh feeder equipment module, cement hopper discharge gate valve is a control module.

Mixer-3 unit consists of three control modules – (i) Water control valve, Mixer-1 discharge pump and (iii) Mixer-2 discharge pump.

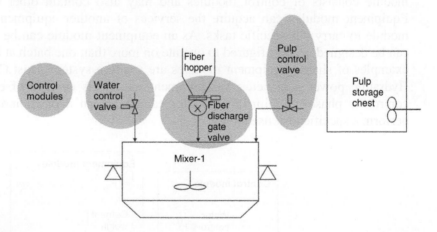

Figure 2.8
Example – identification of control modules

2.3   Define tags

When defining units, it is a good practice to also decide on guidelines or the convention to be followed for tag names. If units are grouped as a class, a scheme for giving consistent names to devices should also be defined. This helps in automatic tags generation during configuration of control systems. Typical tag names for batch process explained in Example 2.1 are given below and indicated in Figure 2.9.

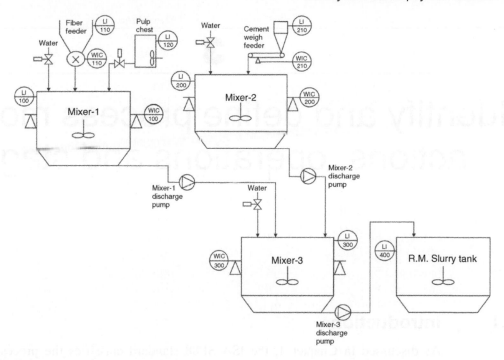

Figure 2.9
Example of defining and giving tag names

2.3.1 Example of typical tag names

Equipment entities are given three-digit tag number. First digit indicates unique unit number and next two digits indicate the unique equipment entity number in that unit.

Three-digit tag number: X YY

Unit equipment entity

E.g. Mixer-1 unit is given unit number 1.
The Mixer-1 is given equipment number 100.
Fiber feeder equipment module is given equipment number 110.
Pulp preparation and storage chest is given equipment number 120.
Similarly, Mixer-2 unit is given unit number 2.
The Mixer-2 is given equipment number 200.
Cement weigh feeder equipment module is given equipment number 210.
Mixer-2 equipment, which rests on load cells for weighing components, its weighing system is given tag name WIC-200. High-level sensor, which indicates maximum level of the Mixer-2 and overflow, is given tag name LI-200.
Cement weigh feeder equipment module is given tag name WIC-210. High-level sensor on Cement hopper in assigned tag name LI-210.

Similarly, tags are assigned to other equipment entities in the process cell.
The tag name convention explained above is a typical one. However, while deciding on tag name convention and guidelines for giving tag names, it is important to ensure that a unique tag number or tag name is given to an equipment entity.

3

Identify and define process models, actions, operations and stages

3.1 Introduction

As discussed in Chapter 1, the ISA SP88 standard classifies the process based on the number of products it manufactures and physical connections as follows.

3.1.1 Single-path process

In a process with single-path structure, the batch passes sequentially in a predefined path from one unit to another unit.

3.1.2 Multi-path process

In a process with multi-path structure, there may be several batches active at a time and the equipment may be of different physical characteristics.

3.1.3 Network process

In a process with network structure, the sequence of the units may be pre-assigned, determined prior to execution of the batch or even during the execution of the batch. In network structure process, an appropriate path is determined at the time of execution, depending on constraints like recipe requirements and equipment capabilities. Control of network process is complex due to the need for allocation of equipment and the arbitration of requests for the equipment.

For a batch process, ISA S88 describes the following models:

- *Physical model*:   To explain physical assets of the enterprise.
- *Process model*:   For sub-division of a batch process.
- *Procedural control model*:   For sub-division of procedural elements of batch process.
- *Procedural equipment mapping model*:   For defining links between the above three models in order to achieve the batch process functionality.
- *Recipe-type model*:   To define the recipe types and relation between different types of recipes.
- *Control activity model*:   To deal with functional activities of a batch system.

In this chapter we will learn to identify and define process model and the relationship between process model and physical model.

3.2 Process model

A batch process can be sub-divided into a hierarchical structure known as process model, as shown in Figure 3.1. Process model consists of four elements, each of them composed of an ordered set of hierarchical sub-elements, as described in the following sections. In the process model, the procedure for making a product does not consider the actual equipment required for performing various process steps.

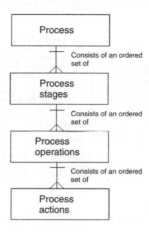

Figure 3.1
Process model

3.2.1 Process

A process is defined as a sequence of biological, chemical and/or physical activities for conversion, storage or transportation of material or energy. The process may be considered to be an ordered set of process stages, organized in a series, and/or in parallel. As such, any one-process stage usually operates independent of the others. It results in a planned sequence of physical and/or chemical changes to the materials being processed. Any process that has finite quantity of material output is known as a batch process.

3.2.2 Process stage

A process stage describes a major function that normally results in a planned sequence of chemical and/or physical changes in the properties of the material involved in the process. Each process stage consists of an ordered set of one or more process operations. Process operations are major processing activities. Process stages operate independent of each other. Process stages may be decomposed further into ordered sets of process operations, such as charge, and process actions, such as heat. Examples of typical process stages are polymerization and drying.

3.2.3 Process operation

A process operation represents a major processing activity. A process operation results in chemical and/or physical changes in the properties of the material being processed. Each process operation can be further sub-divided into an ordered set of one or more process actions that carry out the required processing by the process operation. Examples of typical process operations are preparation of reactor, charge, react, etc.

3.2.4   Process actions

A process operation consists of an ordered set of one or more process actions.

A process action describes several minor processing activities that are combined to make up a process operation. Examples of typical process actions are heat reactor, add catalyst, hold, etc.

Example 3.1:

Process model of fiber–cement sheet manufacturing process is shown in Figure 3.2.

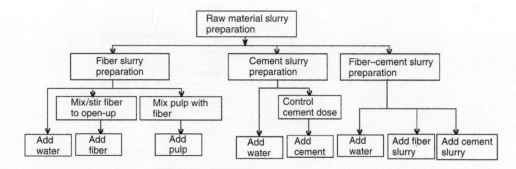

Figure 3.2
Process model of fiber–cement sheet manufacturing process

Example 3.2:

Another example of process model for polymerization process is shown in Figure 3.3.

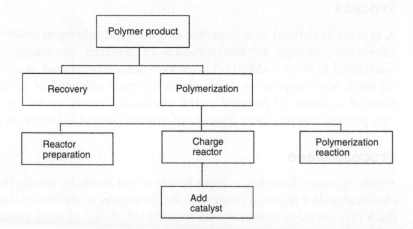

Figure 3.3
Example – process model for polymerization process

3.3   Relationship between process model and physical model

The ISA SP88 Part-1 standard looks at a batch process control from two different points of view. First is the process point of view and second is the equipment point of view. The process point of view is normally considered a chemist's view and is represented by the process model. Whereas the equipment point of view is considered a product engineer's view and is represented by the physical model. The relationship between the process model and the physical model is illustrated in Figure 3.4.

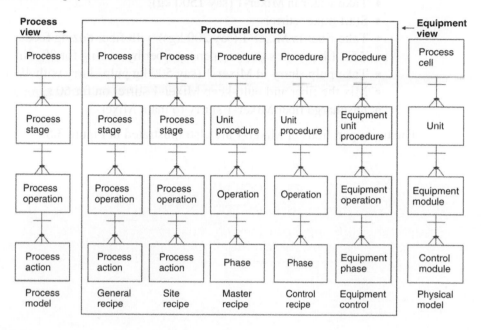

Figure 3.4
Relationship between the process model and the physical model

The process, as defined in the process model, relates to its peer process cell described in the physical model. Similarly, peer-to-peer link or relationship exists between different hierarchical entities of the process model and physical model.

Example 3.3:
Let us consider the example of raw material batch preparation for fiber–cement sheet manufacturing and describe unit operation for Mixer-1 unit as shown in Figure 3.5.

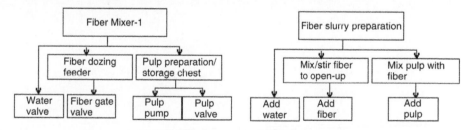

Figure 3.5
Unit operations of mixer

The physical model of the Mixer-1 unit consists of the following:

- Two equipment modules:
  - *Fiber dozing feeder*: Consisting of control module, fiber-feeding gate valve
  - *Pulp storage chest*: Consisting of control modules, pulp pump and pulp control valve.
- One control module – water control valve.

The unit operation of Mixer-1 is:

- Take water in Mixer-1 (say 1500 kg)
- Start mixer stirrer
- Take fiber in Mixer-1 (say 200 kgs) with fiber dozing feeder
- Mix fiber with water to open-up fiber by stirring for 20 min
- Take pulp slurry in Mixer-1 (say 500 kg pulp slurry with 4% concentration)
- Mix the fiber and pulp-keep Mixer-1 stirrer on for 60 s
- Discharge fiber Mixer-1 slurry to main Mixer-3.

The unit operations of Mixer-1 are also illustrated in Figure 3.5.

4

Identify and define procedural models

4.1 Introduction

While discussing the relationship between process model and physical model, we had referred to procedural control, as illustrated in Figure 3.4. Procedural control is the link between process model and physical model. Four levels of recipes and equipment control form a procedural control. Procedural control directs equipment-oriented actions in an ordered sequence to carry out process-oriented tasks.

4.2 Procedural model

The procedural model defines the control that enables the equipment in the physical model to perform a process task. Procedural models are used to describe controls that direct the equipment's actions to perform some process-oriented tasks. The procedural model consists of four elements organized in hierarchy, as shown in Figure 4.1.

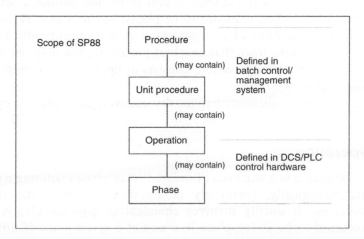

Figure 4.1
Elements of procedural control model

4.2.1   Procedure

A procedure is the strategy for accomplishing or carrying out a major processing action. Or in other words, a procedure is a sequence of unit procedures required to make a batch. It is the highest level in the hierarchy of procedural control model, and it orchestrates the control of the equipment in the process cell. As illustrated in Figure 4.2, a procedure can be further sub-divided and defined in terms of unit procedures, and/or operation, and/or phases. The domain of procedure is process cell.

Example of a procedure is making a batch of fiber–cement slurry for making sheets.

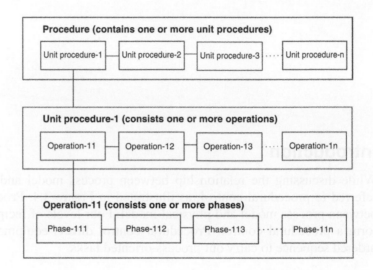

Figure 4.2
Hierarchical sub-division of a procedure

4.2.2   Unit procedure

A unit procedure is a strategy for accomplishing or carrying out a contiguous process within one unit. In other words, unit procedure is a sequence of operations that controls the functions of a single unit. A unit procedure defines a set of related operations that cause a production sequence to be placed within a unit. However, it is possible to have multiple procedures of one process executed concurrently but each in different units. A unit may have more than one unit procedure, but only one unit procedure controls the unit at a time. Unit procedure consists of operations and methods required for initiation, organization and control.

Examples of unit procedure are dry, recover, polymerize or open-up fiber by grinding.

4.2.3   Operation

An operation is a sequence of ordered set of phases defining a major processing sequence that biologically, chemically or physically changes the state of a material being processed. It usually involves chemical or physical changes. Only one operation is presumed to be taking place in a unit at a given point of time. The next operation can begin only on the completion of the first. Typically, an operation controls a portion of the unit functions.

Some examples of operations are fiber slurry preparation and reaction.

4.2.4   Phase

A phase is the smallest element of procedural model that accomplishes a specific process-oriented task. It defines a product-independent processing sequence. A phase performs a simple process function on an equipment module and coordinates the control of control modules. Phases can be executed either sequentially or in parallelly. Phases can be self-terminating without an end command. A phase is the lowest group of process actions. A phase may be divided into steps and transitions according to sequential function charts (SFC) or Grafcet. Design of a phase must take into account exception and safety conditions. Examples of phases are add catalyst, add water to fiber, etc.

Example 4.1:
Procedural model for raw material slurry batch preparation for manufacturing fiber–cement sheets is shown in Figure 4.3.

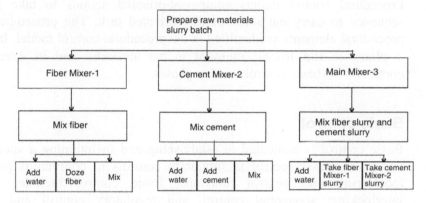

Figure 4.3
Example – procedural model for raw material slurry batch preparation

Example 4.2:
Another example of procedural model for sterilization and fermentation process for manufacturing anti-tuberculosis drug is shown in Figure 4.4. The sterilization process consists of sterilization of the fermentor vessel prior to commencement of fermentation. After sterilization process, the required ingredients ammonia, dextrose, titer, etc., along with accurate quantities of acids and alkali, are added to the vessel and the process of fermentation is carried out.

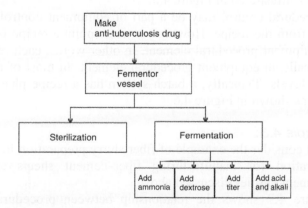

Figure 4.4
Example – procedural model for sterilization and fermentation process

4.3   Concept of equipment entities

An equipment entity is a combination of physical equipment and the equipment control application. Equipment control is equipment-specific functionality that provides the actual control capability.

It is classified into three types of control:

1. Procedural control
2. Basic control
3. Coordination control.

4.3.1   Procedural control

Procedural control directs equipment-oriented actions to take place in an ordered sequence to carry out some process-oriented task. The procedural control consists of procedural elements as identified in the procedural control model. In a procedural model, a phase is the lowest element in the hierarchy and its execution may result in commands to basic control or other phases.

4.3.2   Basic control

Basic control is dedicated to establishing and maintaining a specific state or process condition. Basic control is similar to controls in continuous processes. It performs monitoring and control functions. Basic control includes exception handling, interlocking, sequential control, and regulatory control and alarm annunciation. However for batch processes, there may be additional commands that are needed to be acted upon.

4.3.3   Coordination control

Coordination control directs, initiates and/or modifies the execution of procedural control and utilization of equipment entities. Coordination control includes allocating resources, propagating modes and arbitration of shared resources equipment requests.

The relationship between the procedural control model, process model and physical model is illustrated in Figure 4.5.

Procedural control may be a part of equipment control or passed on via a procedural entity from the recipe. However, at some point a recipe procedural element must map to an equipment procedural element. In other words, each recipe procedural element points to or calls an equipment procedural element. In most of the cases, the mapping is at the phase levels. Typically, a batch system has a recipe phase that references an equipment phase as shown in Figure 4.6.

Example 4.3:
Let us consider the example of fiber slurry preparation in Mixer-1 for raw material batch preparation for manufacturing fiber–cement sheets and illustrate the concept of equipment entities discussed above.

Figure 4.7 shows the relationship between procedural control, process model and physical model for the fiber slurry batch preparation process.

In the fiber Mixer-1, batch control is built around common function entities.

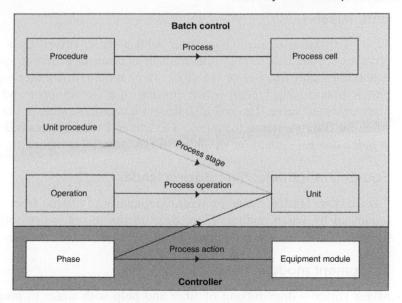

Figure 4.5
Relationship between procedural control, process model and physical model

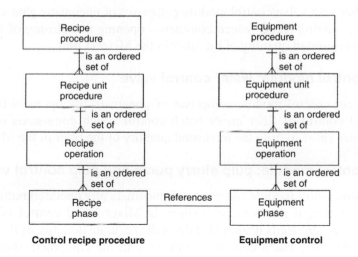

Figure 4.6
Typical procedural mapping

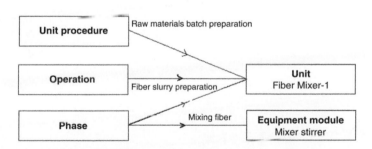

Figure 4.7
Example – integration of procedural control, process model and physical model

Unit: Mixer-1

Mixer-1 consists of a mixing vessel with a stirrer driven by a variable-speed invertor drive to accomplish mixing operation in the unit at different speeds, depending on the consistency and finesses of the fiber slurry required. Another equipment module in the unit is fiber-dosing feeder, which consists of a fiber hopper and a control module fiber discharge gate valve. The unit also houses a control module in the form of a water control valve for filling water in the mixer. The Mixer-1 vessel rests on three load cells connected to batch control system for weighing of raw material components – water, fiber and pulp.

Equipment module: fiber dozing feeder

Fiber dozing feeder, doses the required quantity of ground fiber in fiber Mixer-1 unit, as requested by batch control. Fiber doze in the mixer is controlled based on load cells weighing control module and the fiber gate valve control module.

Equipment module: mixer stirrer

Mixer stirrer – does mixing of fiber and pulp with water to prepare fiber slurry. Mixer time for each raw material component is commanded by the batch control system.

Control module: fiber gate valve

Fiber gate valve control module comprises of pneumatic gate valve and control module of fiber dozing feeder which commands opening and closing of the fiber gate valve to dose the requested quantity of the fiber in the Mixer-1 unit.

Control module: water control valve

Water control module comprises of pneumatically operated butterfly type control valve and control of fiber mixer batch control which commands opening and closing of the water valve to dose the requested quantity of the fiber in the Mixer-1 unit.

Control module: pulp slurry pump and pulp control valve

Pulp control module comprises of pneumatically operated butterfly type control valve and a centrifugal pump to doze slurry in Mixer-1 and control of fiber mixer batch control which commands dozing of pulp in the accurate quantity of the fiber in the Mixer-1 unit.

The basic batch control is built around the equipment modules – fiber mixer stirrer, fiber dozing feeder and control modules water valve, fiber valve, pulp valve and pulp pump.

5

Introduction to recipes

5.1 Introduction

Every batch plant has recipes for the products it produces. Every product recipe consists of a header, equipment requirements, formula and procedures. For example, let us consider a recipe of a cake mix. The recipe header identifies the product as, say chocolate cake. The equipment requirements identify the required processing equipment for making the cake. For the cake mix, the equipment requirements are detailed on the pack, say a large bowl, a mixer, a cake pan, etc. The formula defines the ingredients, such as 1 pack of chocolate cake mix, 2 eggs, 150 ml water, 50 ml edible oil, etc. The procedure defines the processing actions and sequence of execution. For making chocolate cake, first pre-heat the oven to 350 °C, mix eggs with water and oil, add cake mix and blend until smooth, pour into cake pan and bake in oven for 45 min.

Every product produced in a batch plant has a recipe consisting of similar components. For some products, these components may not be obvious, but they do exist. For example, a recipe may be on a piece of paper that the operator uses to produce the batch. The recipe header and formula are defined on the paper. The equipment requirements and procedure is implied, e.g. blend all ingredients together in a blender. In a more automated process that has a control system, the formula values and procedure may be fixed in that control system. In some plants, the products to be produced have the same procedure and only the formula changes. If there are only a few products, then formulas can be stored in the control system and the operator selects the product to be run by pushing a button or turning a selector switch to the desired product. Control system logic is used to switch-in the new formula values. When there are more products than products recipes can be stored in the control system, the capability to create, edit and manage recipe formulas is provided externally. Similarly, a facility that downloads new formulas to the control system may be provided when needed.

While discussing the relationship between process model and physical model, we had referred to various levels of recipes under procedural control, as illustrated in the Figure 3.4. For batch manufacturing of a given product in a process cell, the ISA S88 standard suggests gradual refinement of process model based on four types of recipes. The four types of recipes defined in the standard are general, site, master and control recipes. A recipe contains administrative information, formulas, requirements of the equipment needed and the procedure defining the method to produce the recipe. The procedure is organized according to the procedural control model.

5.1.1   Recipe

A recipe is a data structure that gives production information, formula, procedure, equipment requirements and other related information. A recipe is also an entity that contains a minimum set of information that uniquely defines the manufacturing requirements for a specific product. As shown in Figure 5.1, a recipe consists of five parts that combine together to provide all the information required to produce the product. Each part of the recipe is described below:

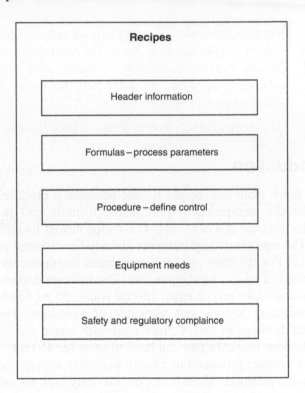

Figure 5.1
Parts of a recipe

Header

The header contains all administrative information for a recipe including the batch name, lot number, product information, version number, author, issue date and the status.

Formula

Formula contains a list of raw materials, quantities of each raw material, operating conditions, etc. required to make the product. In addition, the formula also includes other parameters such as temperature, pressure and process outputs like the recipe's expected production quantity. In the control recipe this is invariably in the form of data in a parameter list.

Product specification

Product specification determines the quality tests and expected results that are performed on the material during the process. The product specification list determines which of the available tests are performed on the material and defines the acceptable range of results.

Equipment requirements

Equipment requirements define a list of the equipments required to produce a product. The equipment relates to the physical model and, for example, identifies equipment and control modules. In the general and site recipes, this list is a general guideline. The master recipe contains a list of equipments that can produce the product. The control recipe defines the specific line that will be used to produce the batch.

Procedure

Procedure defines the process strategy. Procedures enable the progression of batches, as described. At the master recipe level the phases are referred to by name only, but at the control level all the steps and actions are included. General and site recipes include procedures based on the process model. Master and control recipe procedures are defined using the structure defined in the procedural model.

According the ISA SP88 standard, a recipe should be independent of the equipment on which it is executed. Such separation of equipment and procedure lends itself to procedural elements that are transportable between the shared equipment.

Safety and regulatory compliance

Safety and compliance information provide for miscellaneous information and comment as appropriate.

5.2 S88 recipe model

The ISA S88 standard advocates four types of recipes – general, site, master and control recipes. These different types of recipes enable a range of recipe representations varying from the general, which are generic process descriptions, through to the control, which are detailed, unambiguous and equipment-specific. Translation of recipes from the generic to the specific essentially involves a lot of detailing and is quite a complex process. A recipe is defined to be the complete set of information that specifies the control requirements for manufacturing a batch of a particular product. As discussed above, recipes contain five categories of information – header, formulation, equipment, procedures, and safety and compliance information. The information contained in each category should be appropriate to the level of recipe.

The ISA S88 standard Part 1 compliant batch control solutions have provided flexible implementations at many sites worldwide. Master recipes are tailored to equipment trains and parameterized to become control recipes dedicated to a single batch ID. The ISA S88 standard Part 1, recognizes that an enterprise may make a product in slightly different ways at different sites and that a higher-level recipe ultimately defines the product identity or brand characteristics. The gap between the higher-level recipe and the master recipe is often vast. The former recipe comes from chemists in process development, while the latter recipe is produced by process engineers with intimate knowledge of a specific site's equipment and raw materials. A great deal of value must be added to get from the high level to the low level.

The concept of a general recipe is more relevant in case of larger enterprises with multiple sites. The control system suppliers are keen to develop solutions to help bridge this gap as it positions their technology one level higher up the food chain of enterprise IT. The Part 2 of the ISA S88 standard provides guidelines for development of general recipes.

5.3   Types of recipes

As illustrated in Figure 5.2, the S88 standard defines four types of recipes as follows:

5.3.1   General recipe

The general recipe is an enterprise level recipe that describes the requirements to manufacture a product at various sites. It provides a very high-level view of requirements for producing a product and can be used at many sites. General recipe includes general information on required equipment, raw materials and procedure, without regard to the specific production. General recipe serves as the basis for the other recipes. A general recipe is created without specific knowledge of the process cell equipment that will be used to manufacture the product. A corporate chemist usually creates general recipe.

5.3.2   Site recipe

The site recipe is specific to a particular site and describes specific requirements in terms specific to that site. Site recipe is derived from the general recipe by a process engineer and includes information that is site-specific. The site recipe translates the general recipe into a more specific version that allows for the types of equipment and raw materials to be available at the site. The language, units of measurements and raw materials are adjusted to the site. This version of the recipe is designed to be used in many different process cells.

5.3.3   Master recipe

The master recipe is targeted to a specific process cell. Master recipe is derived from a general or a site recipe and accounts for specific information such as equipment arrangements that are necessarily required to make the product. Master recipe is created by the control engineer and is designed to be used on many different lines within the process cell.

A master recipe can also be created, as a standalone entity by the people that have all the information that otherwise would have been included in the general recipe or the site recipe.

5.3.4   Control recipe

The control recipe is a copy of the master recipe, which has been completed or modified with scheduling, operational and equipment information. A control recipe can be viewed as an instantiation of a master recipe. Control recipe is created from the master recipe when a batch is scheduled for production and it defines the manufacture of a single batch of a specific product. This is the most specific version of the recipe. Control recipe is created when a batch is scheduled. The control recipe is executed to make a batch of product. The control recipe includes information that is specific to the batch, the production line on which the batch will be produced and the raw material that will be used. Control recipe is the only type of recipe that is required for production.

The recipe levels are collapsible. A master recipe can be created as a standalone entity without the need for a general recipe or a site recipe. However, it is assumed that the recipe developer has full knowledge of the specific process and its capabilities. The four recipes are gradually refined to the stage where all the required aspects for the execution of the recipe on a certain type of equipment have been considered and taken care of. The general and site recipes are equipment-independent whereas the master and control recipes are equipment-dependent. To distinguish between the equipment-dependent and equipment-independent process steps, different terminologies are used. The terms procedure, unit procedure, operation and phases are used for the equipment-dependent process steps, as illustrated in Figure 5.2.

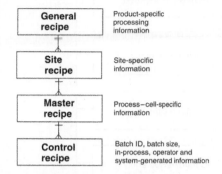

Figure 5.2
Types of recipes

Example 5.1:

An example of a recipe to manufacture toothpaste is illustrated in Figure 5.3.

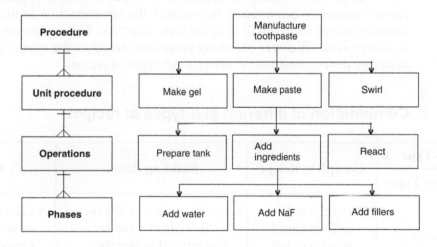

Figure 5.3
Example of a recipe to manufacture toothpaste

5.3.5   Recipe entity

Recipes are organized hierarchically with various categories of information at each level. The construct to represent the tight coupling of the data at a particular level is known as recipe entity. The recipe entity is the fundamental structure in all types of recipes. As defined in the ISA SP88 standard, structure-wise the recipe entity takes the place of the recipe procedural element but it may include any or all of the recipe components like procedural definitions, parameters with values, equipment requirements and other information. A complete entity is considered to be a recipe entity itself. The recipe is built up of lower-level recipe entities or components like unit procedures, operations, etc. When constructing a specific recipe, the components are created using building blocks available from library elements. Recipe sub-types are shown in Figure 5.4.

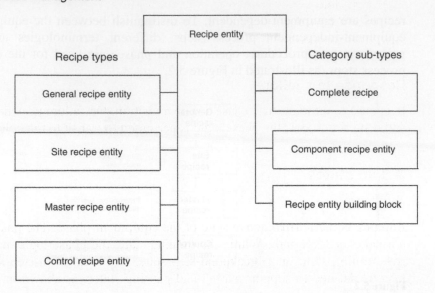

Figure 5.4
Sub-types of recipe entity

The recipe entity concept is applied to all types of recipes – general, site, master and control recipes. When a recipe is executed, the representation in the batch history of the executed recipe entities will have the same structure. The equipment procedural elements also have same structures and many properties similar to the recipe. The following matrix describes the combination of different sub-types of recipes.

5.3.6   Combination of different sub-types of recipes

Sub-Type ▶ Recipe Type	Complete Recipe	Building Block	Component
General or Site	Complete and self-contained general or site recipe	Generic, general or site recipe entity type that can be instantiated in specific recipe or in another building block	A component of a general or site recipe or library element, may be an instantiation of a building block
Master	Complete and self-contained master recipe	Generic master recipe entity type that can be instantiated in specific recipe or in another building block	A component of a master recipe or library element, may be an instantiation of a building block
Control	Complete and self-contained control recipe	Building blocks for control recipes do not exist. Control recipes are modified using master recipe building blocks	A component of a control recipe, may be an instantiation of a master recipe building block

5.3.7 Components of recipe entities

Figure 5.5 shows components that may exist at any level of decomposition of the recipe. The procedural hierarchy is modeled through a recursive relation including those recipe entities that may be made up of lower-level recipe entities. This allows for the definition of the levels defined in the SP88 standard, i.e. procedure, unit procedure, operation and phase. Further, it also allows expandability and collapsibility. Modeling of equipment requirements is done based on the product requirements. The formula is modeled as a set of parameter objects. All levels of recipe decomposition may have parameters including the recipe itself. Other information are represented as a single object class even though other information may have multiple elements and different structures.

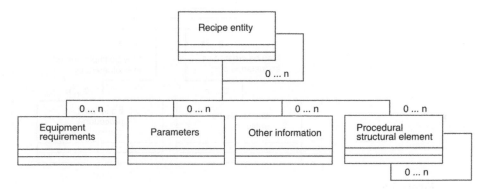

Figure 5.5
Components of recipe entity

The procedural structural elements make up the procedural specification. Lower-level recipe entities may transition between these levels and other language constructs, as described in the standard S88 Part 2.

5.3.8 Recipe entity building blocks

Recipe entity building blocks are the building blocks from which the master recipes are created. Recipe entity building blocks are shown in Figure 5.6. When a recipe entity building block is instantiated in a master recipe as a master recipe entity, it may carry parameters, equipment requirements and other information that may be assigned master recipe specific values. Lower-level contents of the recipe entity building block may be copied into master recipe entities. Referring to the recipe entity building block may also access these same level contents. A recipe entity building block's functionality may be implemented in equipment through equipment procedural elements. This is necessary for the execution of the lowest-level recipe entities – the recipe entities that are intended to be linked to equipment procedural elements.

5.3.9 Recipe parameters

Recipe parameters are variables associated with recipe entities. These variables may be used by equipment procedural elements, scheduling model, or referenced by other parts of recipe like transition criteria, etc. Recipe parameters may be categorized as process inputs, process outputs or process parameters. Parameters may be a set of other parameters as the S88 model supports the concept of structured parameters. Parameters' value attributes may be organized by defining parameter value types. Parameter value types could be based on IEC 61131-3 standard basic data types or application-specific type. Parameter values may be simple values, expressions, or references to parameters defined at the same level or higher levels in the procedural hierarchy. Expression values may include references to other parameters.

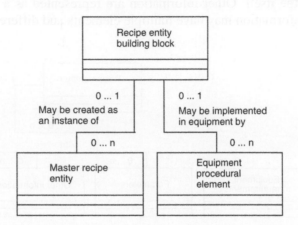

Figure 5.6
Recipe entity building blocks

The ISA S88 Part 1 formula is represented in the data model as recipe parameters. A recipe's formula is a collection of selected parameters to the recipe procedure and may also include parameters defined at the lower levels of the procedural hierarchy.

5.3.10   Recipe procedure mapping with equipment procedure

A recipe procedural element must map to an equipment procedural element at any point of time. A recipe procedural element points to or calls equipment procedural elements. In most cases the mapping is at the phase level. Typically, a batch control system will have a recipe phase which references an equipment phase.

5.3.11   Class-based recipes

A recipe is used to make a product. Whenever possible, master recipe should be defined without any reference to what specific equipment entities it will run on. Equipment is/are allocated only when the control recipe is generated and the recipe procedure is executed.

A recipe procedure does not know how to control the equipment. Control is not a part of the recipe. The recipe procedure is linked to equipment procedures at some level to perform the actual control. This separation of the equipment control from recipe provides flexibility in system implementation. The control recipe procedure can be detailed with references to recipe phases or to high-level recipe procedure, and depending on the equipment control and control recipe, the linking can be done at the phase, operation unit procedure or procedure level.

5.4   Building recipe procedures

As illustrated in Figure 2.5, the physical model of a process is defined from top to bottom, i.e. first we define the process cell, followed by the units, equipment modules and finally control modules. This top to bottom approach helps in enforcing a modular design. While building a recipe procedure, the opposite approach, i.e. down to top, is better. By using a bottom-up approach to design procedural elements, it is much easier to end up with reusable, modular procedures. While defining a physical process, equipment phases are built. These equipment phases now need to be mapped to recipe phases. The recipe phases are defined in batch management software. They are the foundation on which you will build your recipe procedures. After you have defined the recipe phases, it is necessary to look at the product needs. Operations are the highest procedural level that can typically be built with any product independence. Some typical examples of product-independent operations are purge, react and transfer. Each of these could be built in such a way that they would work for any product.

Once you have built product-independent operations, you need to define the remaining operations, which typically contain processing information, such as how raw materials will be loaded for making each specific product, etc. After this, unit procedures that can be used to coordinate the activities of units within the process cell are built. In practice, you can hardly find a procedure for a given product that contains only one unit procedure for each unit. It becomes easier to manage the allocation of resources by sub-dividing the unit procedures, as illustrated in Example 5.2.

Example 5.2:

Let us consider an example of recipe of cellulose pulp refining process consisting of a network of five units, as shown in Figure 5.7. There are two un-refined pulp chests, two refiners and a refined pulp chest. Either of the two un-refined pulp chests can feed to either of the refiner, or either of the refiner can feed refined cellulose pulp in the refined-pulp chest.

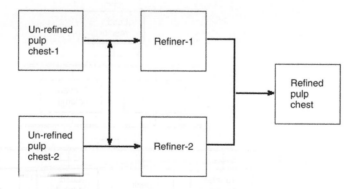

Figure 5.7
Block diagram of an example of recipe process for cellulose pulp refining

For the process in Example 5.2, if you build a recipe procedure that has one unit procedure per unit, the units will by default be allocated to the batch as soon as the batch begins execution. If there is still a batch in the refined-pulp chest, another batch cannot be started in either of the un-refined pulp chests. Most batch engines have mechanisms that can minimize this impact, but it is helpful to build unit procedures that do not acquire the equipment until it is needed. Refer recipe procedure illustrated in Figure 5.8.

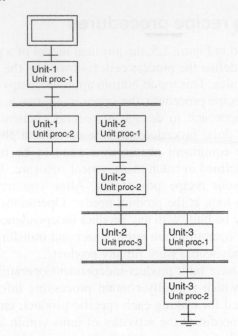

Figure 5.8
Example of building recipe for cellulose pulp refining

Example 5.3:
Consider an example of product recipe procedure illustrated in the Figure 5.9.

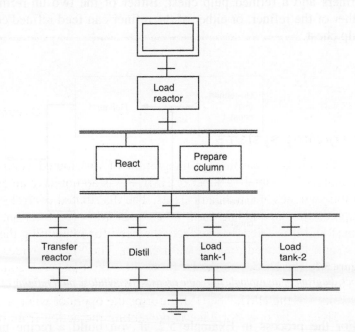

Figure 5.9
Example of product recipe procedure

For the product recipe procedure illustrated in Figure 5.9, the two operations, load reactor and distil, are illustrated in the Figure 5.10.

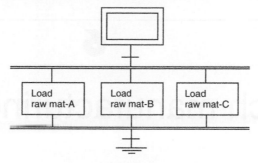

(a) Example operation – load reactor

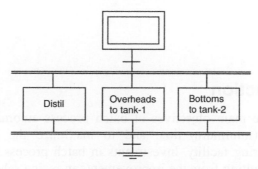

(b) Example operation – distil

Figure 5.10
Operation – load reactor and distils involved in the example product recipe procedure

6

Batch manufacturing basics

6.1   Introduction

One of the main objectives of batch process automation is to manufacture a batch in optimum (minimum) time and to maximize the capacity utilization of the batch manufacturing facility. Investments in batch process automation can be justified by the returns resulting from the improvements in recipe scheduling which are easy to quantify. Various planning and scheduling techniques employing fairly simple to very complex algorithms are used for optimization of batch manufacturing processes in the industry. These planning and scheduling techniques have been developed and used for specific applications, depending on suitability to a particular batch process or an industry. However, these techniques fall short for general use.

6.2   Batch numbering, tracking and reporting

6.2.1   Batch tracking

The main objective of batch tracking is to provide another view of operations organized by specific raw materials, and intermediate and finished products. This is an important capability as the problems that are not apparent in daily operating results can become visible when viewing the operation by product. To make product tracking possible, the transactions in each of the operational areas of receiving, production, inventory, transfers and shipments must be properly time-sequenced.

The batch tracking system provides an audit trail for each batch produced. Batch tracking is also useful in investigations of customer inquiries about a particular batch, catalyst evaluation and revising yields models. Data generally include the target range for each process/quality variable and the actual mean, standard deviation, minimum, maximum, variance, etc. Calculated variables such as yields and energy consumption, etc. are also included. It is necessary to compensate for transport time and actual line-up when recording the variables. For example, when a silo changeover occurs, data from the new batch in the silo must be linked with reactor conditions at the prior time when the resin was made. Extruder conditions need to be similarly lagged. Often there is some blending of resin in a shipped batch. This blending must be recorded in the batch tracking record to permit calculation of weighted average properties. Recycling of resin must also be considered in the expected calculations. Plant batch performance calculations use sophisticated algorithms to satisfactorily handle these processing requirements.

It is desirable to be able to retrieve historical batch information by a variety of criteria. For example, it may be necessary to review all batches of resin-grade M6 produced in the past 6 months. Conversely, it may be of importance to review all batches with deviations from key target variables of more than a pre-selected quantity. A variety of other ad hoc queries are of interest. To provide the necessary flexibility, it is generally advantageous to utilize a relational database (RDBMS) as the data warehouse. The batch tracking system provides the information required to update the master recipes for the plant. Generally, this is an offline function, performed by the technical staff as part of their performance monitoring activities.

Products produced in a batch should be numbered for identification and traceability. As per the ISO 9001 standard for quality systems and other regulatory requirements, products must be identified by suitable means during all stages of production, delivery and installation. For the batch products, where and to the extent that traceability is a specified requirement, a unique identification of individual product or batches are a must.

Quality records, operation logbooks, and all sorts of records and reports related to the batch manufactured should be identified with a unique batch number and maintained for a specific period of time based on regulatory and other requirements.

Batch numbering on products helps in tracking the batch in which it was produced and also in analyzing the product defects and quality-related complaints from customers.

For batch numbering, no specific way of numbering can be prescribed. In most cases it varies from one manufacturer to the other, one product to the other, one industry to another and so on. However, any method of batch numbering should ensure that the batch numbering fulfills the requirement of identification and traceability, and uniqueness.

With the availability of microprocessor-based inkjet printer and other printers, these days it is possible to have the batch number printed or date and online real time printed on the batch products.

6.2.2   Batch records/history

The batch plant must capture all data related to what has been produced in each batch. This may be done in both hard copy form and electronic archives. Batch history requirements are more demanded in the pharmaceutical industry, where all batch events that occurred during production must be captured, including all procedural events such as batch start, hold, restart and complete, process alarms, operator changes, operator comments, recipe procedure execution events, material consumption, material production and trends related to key process variables.

In any batch production process, it is important to have an accurate and complete batch history for various reasons, such as regulatory compliance, to address customer complaints, etc. For example, in the US, Food & Drug Administration (FDA) requirements are one driving factor for pharmaceutical companies. The need for electronic batch records has produced the FDA 21 CFR Part 11 Regulation on Electronic Records and Electronic Signatures that allows batch history to be recorded electronically. This is helping to replace older hard copy records and signatures, which were, previously, the only FDA-acceptable batch records.

Optimizing the plant is another important reason for capturing batch history. The first step for improving the performance of any process is understanding what the process is doing – and collecting a comprehensive history that captures all events facilitates this enhancement. Hard copy is still acceptable for batch records with simple batch processes and in situations where a complete history is not a requirement. But when product procedures are complex and extensive, filling in the hardcopies of batch record can be a time-consuming, inefficient and error-prone exercise.

6.3   Batch planning and scheduling

Every plant executes based on a production schedule. The schedule may come from a planning entity somewhere in the company and it is usually produced on regular schedule such as monthly, weekly or daily. The schedule may be produced in different forms, such as a computer printout, a spreadsheet file or electronically downloaded to the batch execution system. The production schedule received from planning typically holds only addresses of the production of products, and not the other collateral production procedures that must be executed, such as the cleaning procedures that are commonly required in the food and pharmaceutical industries.

The main objective of batch production planning and scheduling is to optimize capacity utilization of batch manufacturing facilities and fulfill customer orders within time. The readily available planning and scheduling packages are based on the business systems for the required data and information. The schedule generated must be readily available to the process operator who interacts with the control systems. The information available with business systems may be used to analyze the manufacturing process to simplify the operations. Various studies have clearly shown that scheduling is successful if implemented at multiple levels and each level includes the most appropriate scheduling function. Scheduling information requirements, constraints and timeliness vary with the enterprise, site, area and process cell needs. The benefits of improved planning and scheduling accumulate from a number of tangible and intangible areas. A major quantitative area is the ability to respond faster to requests for high-profit special orders. The flexibility inherent in the planning system permits these orders to be inserted in the production schedule. A second major benefit area is improved feedstock utilization. In some cases, alternate feedstocks of lower quality and cost can be evaluated and used to meet standard production requirements.

6.3.1   Production scheduling

The term scheduling is used for actual assignment of resources to production tasks and sequencing and timing of these operations. A scheduling decision usually is based on medium-term and long-term production planning. In production planning, decisions are taken about the products to be made, quantities to be manufactured, production sites where the manufacturing is to be carried out, over a number of months with a resolution time of one week to a month. During planning, the bottleneck resources are considered in an aggregated manner. Account of weekly or monthly production capacities of sites for the products is considered. Production planning is normally done based on linear models that include demands, capacities, prices, production and transportation costs, capacities, and costs of storage of intermediate and finished products.

Production scheduling is concerned with setting the sequence, quantity, and quality of actual grade production. Specific shipment dates and current inventories are used to set these schedules. Equipments, particularly reactors, have limitations and the interactions among common equipment place constraints on the schedule. It is generally desirable to minimize grade switching and shutdown periods. Many plants have been scheduled in the past with manual procedures, perhaps augmented by spreadsheets. However, a new level of technology is now available to address scheduling needs. These scheduling tools use intuitive user graphical interfaces that allow flexible representation of plant flow sheets combined with discrete event simulators to aid the scheduler. Heuristics concerning operating policies are often incorporated to speed exploration of alternate policies. Figure 6.1 illustrates the data flow for a typical modern scheduling system. With these scheduling systems, the utilization of multiple lines (for example – extruders, etc.) is optimized to minimize downgrades. Raw material requirements are projected along with inventories of different grades and shipments and represented as graphs, charts, and lines, often color-coded, on the user interface. Equipment constraints are explicitly considered. Scheduling systems have sometimes been installed 'stand-alone' requiring manual entry of all data. However, experience suggests that these systems will eventually fall into disuse, as users tire of the routine data input requirements. Much more effective and useful systems can be gained from integration of the scheduling system with the remainder of the information system with automatic updating of current conditions and future requirements.

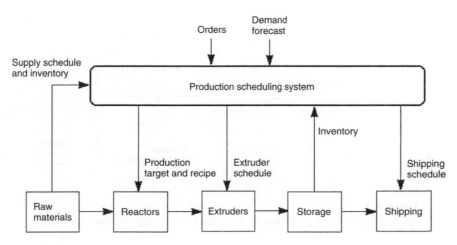

Figure 6.1
Typical scheduling system

Many control systems offer a batch engine, a piece of software that coordinates the development and maintenance of recipes, manages the production of the batch according to the recipe and collects data against a unique batch ID. The ISA's S88.01 standard (IEC 61512-1) defines models and terminology that have enabled the batch community to communicate with each other; most batch engines adhere to this model. S88.01 does identify production planning and scheduling as a key activity, but most systems offer only a simple queue of batches, which are put through the plant in turn.

Most users need to be able to alter this list; modifying it in light of new orders, raw material availability, equipment failure, etc. To cater to this need vendors are now supplying packages allowing the creation of a series, or campaign, of batches. All batches in the campaign can be manipulated as one; processed in a different train, use alternative raw materials, made in a different order, etc. In conjunction with finite-capacity scheduling techniques, these packages are able to add the value of an experienced plant manager who knows how to get the best out of the plant.

Successful implementations are those where the planning and scheduling layer allows the flexibility needed by the individual user. There is no standard such as S88 to help us explain ourselves, and it is an area where an apparently simple requirement can translate into several man-years of effort on the part of both vendor and client.

Planning and scheduling systems are used to reap the benefits such as cost savings and improved customer responsiveness. When it comes to implementing the plans and schedules, the person involved, the domain and the time horizon are different with the planning vs scheduling. The scheduling differs from planning in the near term sequencing and alignment of manufacturing resources to orders. Planning typically involves a much longer time frame and provides greater visibility to strategic goals. Planning may even help in coordinating costs and activities between manufacturing sites. Relationship between planning and scheduling with respect to space and time domain is shown in Figure 6.2.

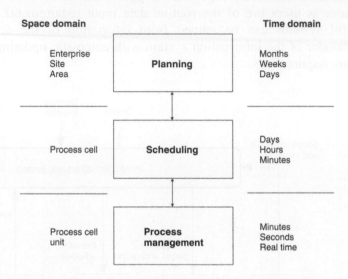

Figure 6.2
Relationship between planning and scheduling w.r.t. space and time domain

It is seen that scheduling often duplicates the activities of planning and process management without any significant value addition. Typically, if there is a planning system that effectively addresses the needs of planners and master schedulers, then the recipe schedule (which is process cell focused and contains product specific information) can become a part of the plan. The time horizon for a recipe schedule depends on the process speed and may be measured in days, hours or minutes. The recipe schedule is based on resources requirement for the process cell, possible paths and sharing of equipment. An effective recipe schedule must be capable of being adjusted in real time to take care of differences in resource assumptions, time projections or other anticipated factors considered while preparing the schedule.

6.3.2   Recipe schedule

Figure 6.3 shows S88.01 control activity model. Production planning and scheduling is regarded as a high-level activity similar to production information management and recipe management. The S88 standard recognizes that these high activities are assigned to process management within a process cell. The recipe focused, combined scheduling and process management activity is referred to as recipe schedule. Inputs for the recipe schedule come from other schedules, master recipes and resource databases. Recipe schedule is connected to higher-level planning and scheduling and is integrated with process management activities.

In the standard S88.02, Batch Control Part 2: Data structures and guidelines for languages, the following properties of the recipe schedule are depicted with help of an object model:

- Batch ID
- Customer requirements

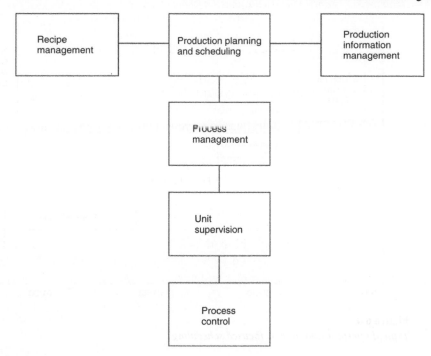

Figure 6.3
Control activity model

- Disposition
- Equipment allowed
- Lot ID
- Master recipe name
- Materials permitted
- Packaging requirements
- Product name
- Product quantity
- Mode of operation
- Start and end time.

6.3.3 Operator and process-focused scheduling

The recipe scheduling must give a view of the status of working recipe sequencing and alignment at-a-glance. The view should be available at remote locations via intranet and should have full timeline scrolling capabilities. The operator should be able to modify the schedule without affecting the background real-time execution. Typical operator and process-focused scheduling should look similar to a recipe schedule shown in Figure 6.4.

Scheduling with an explorer style view of control architecture simplifies the schedule development by using the drag-and-drop techniques to automatically define the execution properties. Visual development environments offer benefits of hiding the real world complexities in graphical representations and making the scheduling simple.

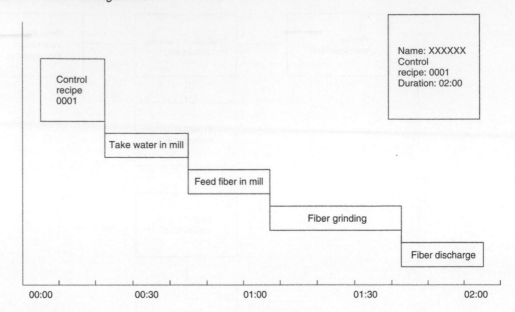

Figure 6.4
Typical operator and process focused scheduling

6.3.4   Integrating schedule to control systems

Planning and scheduling applications include many features supporting optimization, leveling, feasibility and simulation activities. These applications should not duplicate the features and capabilities of batch creation and recipe management and create the redundancy. As the benefits derived from batch execution and recipe management depend on tight integration with process control–user interface, the batch execution systems will ensure that the plant equipment and devices behave properly during normal as well as process upsets with robust failure recovery and error handling.

The scheduler must provide drill-down view of the executing recipes for visibility of parallel unit alignment and synchronization of networked process structures to maximize the equipment utilization. A typical drill-down recipe view from the scheduler is shown in Figure 6.5.

Multi-grade, multi-product batch execution provides formulations and dynamic unit sets to ensure that the process cells throughputs are optimized. As the batch execution provides advanced features for effectively managing both the process structure and the product mix, these features should not be duplicated in a scheduler.

6.3.5   Integrating schedule to planning systems

The recipe management systems allow connectivity to planning systems for materials properties and formula parameters, enabling the planning system to manage orders based on the plant capacity and manufacturing resources. The production information in open journal form meets the requirement of electronic batch recording for the planning system.

To integrate scheduled recipes with the planning system, OLE for process control or OPC connectivity is the best implementation as object methods, properties and events are all that are needed to effect production responsiveness to the customer's orders. Events can be represented visually from scheduler timeline to the current time bar until triggered. Unlinked events can float around the current time bar or may be linked and hover, once the predecessor is finished. A typical visual display of scheduled events is shown in Figure 6.6.

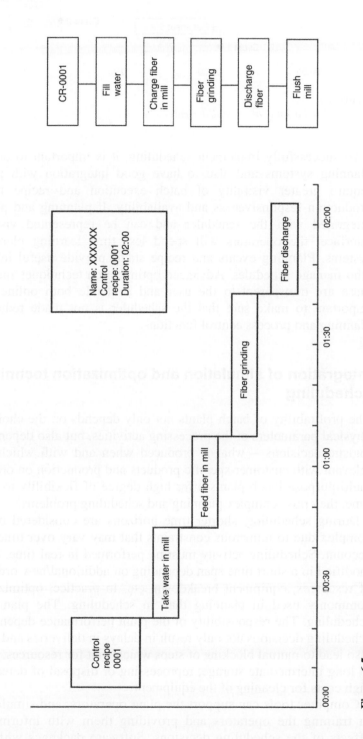

Figure 6.5
Typical drill-down recipe view

planning systems tend to have good integration with precise control, operation input focused flexibility of batch execution and recipe management, to minimize production disruptions and to eliminate diversity and recipe control systems are beneficial. The scheduler and plant personnel usually know the schedule guarantees that a given process will result in a timely and precise control systems. Plant events and recipe also provide useful information to the operators who manage the scheduler and recipe information that are applied effectively. If these are managed in the standard plant data online, real time, plants is important to notice that that the data necessary data is of such excellent that a plant-wide process control has been.

6.3.5 Integration of simulation and optimization techniques to support scheduling

The profitability of batch plants not only depends on the choice of processes, parts and physical parameters, or processing quantities, but also depends to a larger extent on the business decisions of what is produced when and with which equipment. This is more relevant with respect to the products and production on order. Though multi-purpose batch plants have a high degree of flexibility in production demands can turn the plant into a complex planning and scheduling problem.

During scheduling, short-term optimums are calculated only for periods that are completed due to numerous constraints that may vary over time and have to be taken into account. Scheduling activity must be performed in real time by the schedule should be specific enough in time span depending on additional flow orders or orders in production of recipes or equipment breaks down. Also, in practice optimizations are based on many assumptions that it is hard to fully apply to scheduling. The plant operator always do the scheduling. The experienced operator information depends on the operators as bad as skilled the operators the only result of calibrated control and higher priorities given to the normal discipline of orders which necessary represents organization of product quality are more important the closing processes and impact of deterioration of material and other ways for checking of the equipment.

Computer tools can give the plant personnel and simulation of the plant can help in training the operators and providing them with information to check long-term relieves of the scheduling decisions. Software packages with graphical planning tools provide support for adherence to resource and timing of the operations. These tools should be directly linked with a simulator so that behavior of the plant can be.

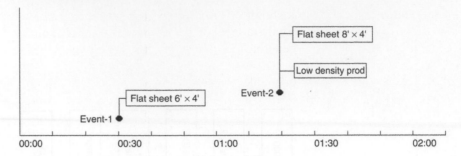

Figure 6.6
Typical visual display of scheduled events

To successfully implement scheduling, it is important to ensure good connectivity to planning systems and also to have good integration with process control. Operators require greater visibility of batch execution and recipe management to maximize production responsiveness and availability. If planning and process control systems are integrated with the scheduler and can be represented visually from the scheduler interface, the operators will spend less time learning planning and process control systems. Planning events and recipe status provide useful information to the operators who manage schedules. Advanced optimization techniques may be applied effectively if these are transparent to the user and available both online and offline. Finally it is important to make sure that the scheduler is not made redundant with duplication of planning and process control functions.

6.3.6   Integration of simulation and optimization techniques to support scheduling

The profitability of batch plants not only depends on the choice of processes, paths and physical parameters of the processing activities, but also depends to a larger extent on the logistic decisions – what is produced when and with which equipment. This is more relevant with customer-specific products and production on order. Though multi-product, multi-purpose batch plants offer high degree of flexibility to meet specific demands on time, they pose complex planning and scheduling problems.

During scheduling, shorter time horizons are considered but the problems are more complex due to numerous constraints that may vary over time and have to be taken into account. Scheduling activity must be performed in real time, i.e. the schedule should be modified in a short time span depending on additional/new orders, changes in availability of resources, equipment breakdown, etc. In practice, optimization techniques are more commonly used in planning than in scheduling. The plant operators always do the scheduling. The responsibility of the plant performance depends on the operators, as bad scheduling decisions not only result in delays in deliveries and higher inventories but may also lead to mutual blocking of steps which wait for resources, degradation of product due to long intermediate storage, reprocessing or disposal of defective/rejected material, and high costs for cleaning of the equipment.

Computer tools can support the plant operators, and simulation of the plant can help in training the operators and providing them with information to check long-term effects of the scheduling decisions. Software packages with graphical planning tools provide support for allocation of resources and timing of the operations. These tools should be directly linked with a simulator so that behavior of the plant can be visualized. These tools should also be integrated with the process control systems for automatic update of actual state of the plant and recipes. Graphical planning tools may include heuristics to generate a basic schedule, which may then be modified by the operators.

Optimization techniques like constraints programming, mathematical programming (either linear or non-linear) and generic algorithms, though not used commonly, are available for determining the optimal schedules. Generic algorithms can be used in conjunction with simulation of a model of desired degree of accuracy to evaluate the admissibility and the quality of the schedules without using any special mathematical model. However, generic algorithms should not be simply put on top of a simulation model as it may be insufficient, as simulation models have several constraints that may not be taken into account while generating the solutions. Besides, the simulation model requires detailing and simulation times are too long.

Mathematical programs are based on certain degree of abstraction of the problem, due to the modeling effort and exponential increase in computation time with the size of the model. Hence, they cannot represent all the constraints that exist in reality. However, after the solution has been determined, a simulation of complete model is advisable to ensure the feasibility of the schedule. Therefore, for multi-product or multi-purpose batch plants it is suggested to use both the optimization techniques and the simulation to obtain a near-optimal feasible schedule. But as it is undesirable and error-prone to make two different models of a batch plant for optimization and simulation, a general-purpose optimization tool may be coupled to a simulator or the simulation is just visualization of the result of the scheduling, for example, a PERT chart derived using linear program. Normally a simulation model is more comprehensive than a scheduling model, but a scheduling model cannot be derived from a simulation model. To generate a scheduling model, aggregation is required rather than just simplification of simulation model. Scheduling algorithms require explicit representation of constraints whereas simulation model implicitly contains constraints. For example, Figure 6.7 shows a pulp refining plant model, in which the four pulp storage tanks are connected in series (single-path process) and transfer of pulp between them is restricted.

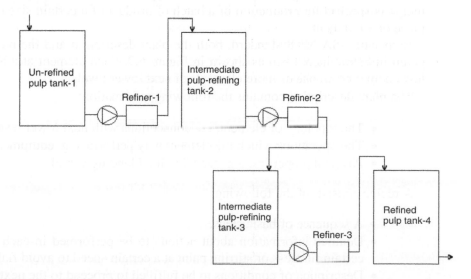

Figure 6.7
Example – single-path pulp-refining process

6.3.7   Core simulation model

There are three key structuring elements in the reference model as described in the ISA standard SP88 Part 1. These three elements are process, plant and recipe description. The process model is sub-divided in plant description and recipe description as shown in Figure 6.8.

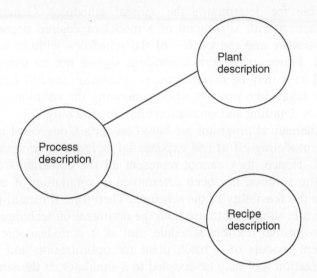

Figure 6.8
Key structuring elements of reference model

The separation between the plant and recipes indicates the flexibility of multi-purpose batch plants. The master recipe describes the procedural steps required to produce a product without identifying these steps with specific plant equipment. The plant description indicates the physical structure of the plant and the elementary functions, which the elements of the plant may perform. Equipment pieces are assigned to operations in the recipe when the master recipe is refined to be a control recipe. Control recipe is specific for production of a batch of product of a certain size using the assigned piece of equipment.

As per the ISA SP88 standard, both the plant description and the recipe description is given in hierarchical form as shown in Figure 6.9. Each element at a higher hierarchical level may contain one or more elements of next lower levels.

The plant description contains the following information:

- The topology of element, i.e. connections with next lower level elements
- The operations which the element may perform, e.g. equipment phases
- Physical properties, e.g. capacity of a blending vessel.

A recipe consists of the following:

- A sequence of basic steps.
- Detailed information about actions to be performed in each step, e.g. pulp at certain fineness or stirring paint at a certain speed to avoid flakes formation.
- Description of conditions to be fulfilled to proceed to the next step, e.g. transfer pulp to refined pulp storage tank only after the pulp is refined to the required degree of fineness.

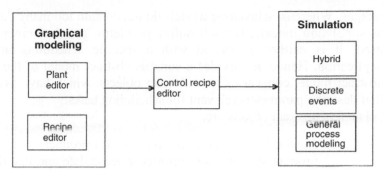

Figure 6.9
Overview of a typical batch simulation package

Different types of recipes are defined. A general recipe is that which contains just the process production route. A site recipe considers equipment at a particular site into account. The master recipe contains all steps involved in the production but no specific reference to specific pieces of equipment. A control recipe which is further refined from the master recipe refers to specific pieces of equipment.

The core simulation model consists of the plant model and the master recipe. The plant model can be formulated on different levels of abstraction. Technical functions provided by each equipment and the resources required are reference in the recipes.

6.3.8   Simulation

In a batch simulation package, the execution of a set of control recipes on a given plant can be simulated. First, the control recipes are generated from the master recipe by allocation of resources, i.e. coupling of recipe phases to equipment phases and deciding on the production sequences and starting times. The scheduling algorithms perform tasks automatically such that optimal results are obtained. Simulation can be used for different tasks during the life cycle of a plant such as generation of plant layout, finding bottlenecks, developing recipes, training the operators, etc. Each of these tasks has different requirements.

A recipe-driven plant can be described by switched systems of algebraic and differential equations representing the transfer of heat and matter in the plant and the chemical reactions and separations. The batch plants have a characteristic that a large number of operations in the plant can be performed in parallel. Whenever a transition condition is fulfilled or a physical change of dynamics takes place, the describing equation changes not only in terms of parameters but also structurally. Hence, the rigorous simulation of recipe-driven batch plants must be based on hybrid models of system's dynamics, where the logical system interacts with a structure-variable continuous dynamical system. The batch simulation packages contain simulators of hybrid systems in which the state vector can be re-configured, based on the evolution of the production process. This gives detailed and accurate results. Overview of a typical batch simulation package is shown in Figure 6.9.

6.3.9   The scheduling model

For automatic scheduling, it is most important that the problem be modeled adequately. Models used for scheduling are quite different from simulation models. In case of simulation, all the degrees of freedom are removed from the description of the plant and recipes. Whereas, scheduling models do not contain too many details and are not accurate as simulation models. For scheduling problems, activity-centered modeling approach is used. It is neither associated with a specific scheduling area nor with a specific application. Hence, it provides a suitable abstract model of the scheduling domain. The abstract model consists of five abstract objects, which have associated set of properties that describe parameters relevant for scheduling tasks.

The five abstract objects are:

1. *Demands*:   Demands are requests for product or services. Demand consists of product specification, priority, release date and due date. Demands are orders from customers or intermediates that require a certain amount of material to be produced and delivered before due date. Sets of activities which are able to produce the requested product or provide services are associated with each demand.

2. *Activities*:   Activities represent processes, which require some capacity of set of resources over a certain period of time. An activity has a starting time, a duration and a set of resources which are allocated during the period the activity is executed.

3. *Resources*:   Resources are necessary for executing the activities. A resource has a capacity which the number of activities can be allowed to use at the same time.

4. *Constraints*:   Constraints impose restrictions that must hold for a specified set of resources or a set of activities or for a mixed set of specific activities on specific resources. Constraints may relate activities to one another to impose sequencing constraints.

5. *Cost function*:   A cost function measures the quality of a schedule in terms of delays in fulfilling the demands and the amount of inventory held due to lack of demand.

Scheduling is the allocation of resources to activities to fulfill a given set of demands with constraints on use of resources and restrictions on execution of activities. The main objective of scheduling is to minimize cost function or to attain a Pareto-optimum for a number of cost functions.

In case of batch process plants, there exist a lot of constraints like shared resources, limited connectivity of resources, scaleable batch size, sequence-dependent changeover and cleaning operations, limited storage time, etc. There are additional constraints also like limited buffer capacities.

The activity-centered modeling viewpoint provides a generic structure and is commonly used for describing scheduling domains. Its extension is used for batch scheduling problems as a template to which the information of core model and additional information such as demands and cost function is mapped. The mapping can be executed in different ways depending on different levels as the activities may be defined. The different ways of mapping the procedural elements of the control recipes to activities help in building the model with different dimensions and reduce the complexity of the model. Complexity can be further reduced by skipping particular constraints of a specific type, which is made possible by the explicit presence of the constraints. The resulting model contains all the activities that have to be preformed. These activities are either necessary to fulfill demands or they are constraints. Hence, the activity-centered model defines the structure of the problem and the dimension of its representation. The most appropriate scheduling algorithm should be used for solving specific problems, and the use of a particular approach should not be a limitation.

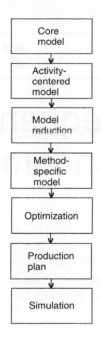

Figure 6.10
Scheduling framework

A typical scheduling framework, as shown in Figure 6.10, with different simulation and scheduling tools, provides a flexible approach to handle scheduling problems related to batch process plants. Models with different dimensions and simplifications can be generated. These models can be easily translated into different representations to select an appropriate model and scheduling tool. The scheduling framework also provides the possibility to simulate the obtained schedule. This permits validation of a schedule and enhances the accuracy of starting time of the control recipes.

7

Batch and sequence programming fundamentals

7.1   Introduction

As discussed in Chapter 4, a procedure is a strategy for carrying out a process to manufacture a product. For mapping procedural elements and equipment there are many alternatives. In a process cell, there may be units, equipment modules and control modules. An equipment phase may exist within a unit or within an equipment module, and may act on equipment or control modules. There are two types of procedural elements – recipe and equipment. Each of the hierarchical levels, procedure, unit procedure, operation and phase, can be either a recipe procedural element or an equipment procedural element. A recipe procedural element is independent of the equipment on which it executes, but an equipment procedural element is specific to a piece of equipment.

7.2   Techniques for batch control elements

In batch management software for recipe execution, all the procedural elements in the batch engine are recipe procedural elements. All the procedural elements in the control system are equipment procedural elements. The recipe procedural elements must be linked with equipment procedural elements. The ISA S88 standard allows procedural elements to be linked across any level. Most of the batch management software systems link recipe procedural elements and equipment procedural elements at the phase level, as illustrated in Figure 4.6. In such cases, the equipment phase is the only procedural element that is specific to the equipment. The equipment phases typically run in the PLC/DCS systems. The recipe phase is run in batch engine.

7.2.1   Modes and states

Each equipment procedural element has distinct number of modes of operation, which need to be defined. Typical modes of operation are auto, semi-auto and manual. In auto mode, the equipment procedural is executed by either a higher-level procedural element or a recipe procedural element. In manual mode, an operator acquires the phase and executes it manually. In manual mode, the procedural element still runs the process, but the operator decides what do. Such an operation in manual mode is different from manually operating the process, like manually starting and stopping pumps or opening and closing valves.

A procedural element can occupy a number of states. Figure 7.1 illustrates a typical state diagram for a phase. An equipment procedural element may have all or some of these states apart from some additional states not illustrated in the Figure 7.1. All the states must be defined and the methods of moving between these states must also be defined. The real task of the phase is performed in the running state and remaining states are for exception handling task.

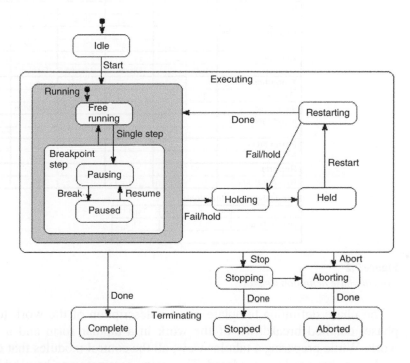

Figure 7.1
Typical state diagram for a phase

7.3   Implementation

7.3.1   Defining an equipment phase

While defining an equipment phase, steps, transitions, parameters, alarms, operator messages, permissive, interlocks and communication with other entities are to be considered. A phase is a sequence of steps and transitions used to perform a process-oriented task. Each step can perform one or more functions such as starting a pump, opening a valve, etc., and between each step is a transition. The transition defines the conditions that must be true for moving between the steps. At the phase level, steps are not performed in parallel. However, the steps need not necessarily be performed in order. Within a phase, there can be decision branches and loops if required. Work to be performed during a phase can be defined in many different ways using structured text, tables, SFCs, graf chart, etc. A combination of text, tables and SFCs is commonly used.

7.3.2   Phase definition template

A typical phase definition template used to build a phase is illustrated in Figure 7.2.

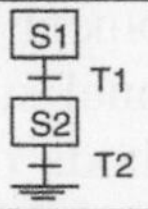

State	Running		Stopping	Aborting	Holding	Restarting
Step	S1	S2	S1	S1	S1	S1
Requests						
Acquire valve-1						
Internal modules						
Feed valve	Open		Close	Close	Close	
External modules						
Valve-1		Open	Close	Close	Close	
Alarms						
Timeout	Enable		Disable	Disable	Disable	
Transitions						
Time > P1	T1					
Valve closed		T2	Complete	Complete	Complete	
Messages						
Parameters						
P1 – Time						
Reports						
R1 – Actual time						

Permissives and Interlocks

Condidtion	Type
Level not high	11

Figure 7.2
Typical phase definition template

The phase definition template gives a description of the work to be performed by the phase, detailed breakdown of the work in a tabular form and a graphical view of the work. While preparing a tabular view, all the control modules that can be manipulated for each step, have to be considered. The recipe parameters to be used have to be defined and also how they have to be used. The tabular view also shows how the alarms will be managed. All alarms related to equipment module or a unit are shown and can be manipulated in any step. Alarms can be enabled or disabled and their priority can be changed. Each operator message is defined and associated with a specific step. The permissions that allow the phase to operate, and the interlocks which cause the phase to fail to a hold state are also defined.

For each of the active states, what the equipment phase will do is defined, including the following:

- The communication the equipment phase will send to a batch management system
- The messages the equipment phase will send to internal modules – equipment modules or control modules
- Alarm management
- Transitions
- Messages the equipment phase will send to the operators
- Parameters equipment phase will receive
- Parameters equipment phase need to send to the batch history for reports

- Interlocks that will cause the phase to fail to a hold state
- Permission that will allow the equipment phase to operate.

The equipment modules and control modules referred to as internal modules are part of the unit or the equipment module. An equipment phase does not communicate directly with output devices. The equipment phase communicates through another module. For example, to open a valve the equipment phase will send a message to a valve control module that will in turn energize the output. The valve control module is an object that owns all the data and methods associated with the valve. This allows different phases to acquire the control module at different times through allocation and arbitration, and send messages to the valve control module.

7.3.3 Transfers

A transfer is a unique type of procedural control. During a transfer there is a need for coordination across units. Transfers can be handled in two ways:

1. To have a transfer out from one unit at one end and a transfer in to another unit at the other end is shown in Figure 7.3. Each unit controls its own resources – i.e. pumps, valves, etc. to achieve the transfer. The units communicate through the batch management system to coordinate the transfer.

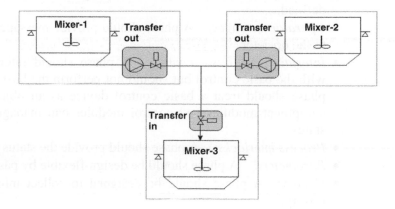

Figure 7.3
Phase coordination transfer pair

2. To have a supply equipment module that has a phase to control the transfer. This can be done with one phase for feed system feeding two mixers as illustrated in Figure 7.4. In case of complex processes, for each combination of sending and receiving units, a phase is required.

7.3.4 Defining phase structure

A phase is the lowest level of procedural control. A phase performs process actions and requires parameters and procedural logic. The phase should have a modular and pre-determined structure, which is easy for implementation and documentation.

While defining the phase structure, following points should be taken into consideration:

- *Command*:   Define and follow a set of commands to phases.
- *State*:   Use predefined states consistently.

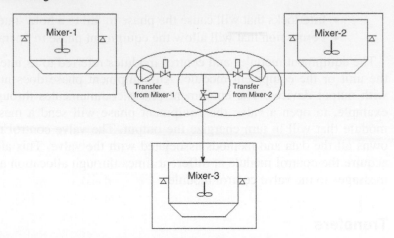

Figure 7.4
Equipment module for transfer control

- *Mode*: Define modes for phases and how phases will respond in each mode.
- *Exception handling*: In case the phase has to take care of exception handling for errors, exception handling and recovery mechanism should be defined.
- *Operator messages*:   A phase should provide messages about phase execution to the operator.
- *Interface with basic control*:   A phase should interface and communicate with the basic control but should not perform the basic control functions. The phase should treat a basic control device as an object. In this manner, the equipment modules and control modules can manage their own modes and states.
- *Process interlocks*:   A phase should provide the status of process interlocks.
- *Parameters*:   A phase should be design-flexible by passing parameters.
- *History*:   A phase should be designed to collect information for production history.
- *Alarms*:   A phase should not generate the alarms directly. It should treat alarms as objects. Based on the alarm conditions, a phase can take corrective actions or perform exception handling.
- *Programming*:   A phase should be designed as a state machine. This helps in selectively enabling alarms and devices.

Example 7.1:

Let us consider an example of transfer equipment module as shown in Figure 7.5. The transfer equipment control module controls the phases necessary to transfer the material out of the vessel and recycle it back to the vessel.

- *Transfer out phase*:   During the transfer out phase, the material from the vessel is transferred to another vessel in another unit. The transfer out phase is illustrated in Figure 7.6.
- *Recycle phase*:   During the recycle phase, the material is recycled back in the vessel. The recycle phase is illustrated in Figure 7.7.

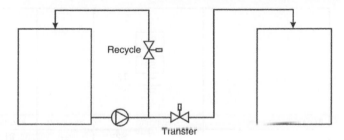

Figure 7.5
Vessel transfer equipment module

	S1	S2	S3	Starting	Aborting	Holding	Restarting
	S1 T1 S2 T2 S3 T3						
State	Running			Starting	Aborting	Holding	Restarting
Step	S1	S2	S3	S1	S1	S1	S1
Requests							
Unit ready to receive	01						
Transfer complete							
Internal modules			02				
Discharge pump		Start	Stop	Stop	Stop	Stop	
Transfer valve		Open	Close	Close	Close	Close	
External modules							
Alarms							
Transitions							
Unit ready to receive	T1						
Qty transfered > P1		T2					
Valve close/Pump stop			T3	Complete	Complete	Complete	
Messages							
Parameters							
P1 amount to transfer							
Reports							
R1 actual qty transfered							

Permissives and interlocks

Condition	Type

Figure 7.6
Transfer out phase

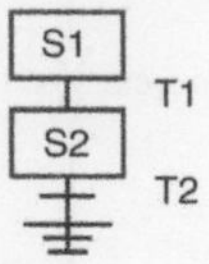

	S1		Stopping	Aborting	Holding	Restarting

State	Running		Stopping	Aborting	Holding	Restarting
Step	S1	S2	S1	S1	S1	S1
Requests						
Internal modules						
Discharge pump	Start	Stop	Stop	Stop	Stop	
Transfer valve	Open	Close	Close	Close	Close	
External modules						
Alarms						
Transitions						
P1 > 0 AND time > T1	T1					
Valve close/Pump stop		T2	Complete	Complete	Complete	
Messages						
Parameters						
P1 recycle time						
Reports						

Permissives and Interlocks

Condition	Type

Figure 7.7
Recycle phase

7.4   Interaction with continuous process sections

As discussed, for a batch process, batch management software is used for recipe execution; all the procedural elements in the batch engine are recipe procedural elements, which must be linked with equipment procedural elements. Most batch management software systems link recipe procedural elements and equipment procedural elements at the phase level.

In case of continuous processes, no batch management system is used. There is no batch engine and no recipe procedural elements. All the procedural control in case of continuous processes is done with equipment procedural elements as shown in Figure 7.8.

Equipment control

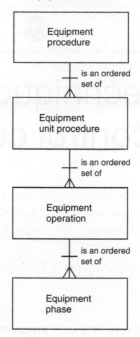

Figure 7.8
Procedural control for a continuous process

8

Practical techniques in sequence control design

8.1 Introduction

There are many different ways of programming PLCs/DCS and a computer. However, there are some standard programming techniques for structured programming, which make program codes more portable, understandable and reusable within and across platforms. In this chapter we will discuss some practical techniques for programming sequence control.

8.2 Programming PLCs/DCS

The International Electrotechnical Commission (IEC) 1131-3 standard on Programmable Controllers – Programming languages specifies the syntax and semantics of a unified suite of programming languages for programmable controllers. The standard specifies two textual languages, namely – instruction list (IL) and structured text (ST), and two graphical languages, namely ladder diagram (LD) and function block diagram (FBD). The sequential function chart (SFC) elements are defined for structuring the internal organization of programmable controller programs and function blocks. Also, configuration elements are defined which support the installation of programmable controller program into programmable controller systems.

8.2.1 Ladder diagram (LD)

Ladder diagram is a conventional form of programming the programmable logic controllers (PLCs). Ladder logic was originally developed as replacement for electrical relays in control circuits and motor starter circuits. PLCs were used to implement control circuits similar to wired-control circuits with electrical relays. Over time, many advanced control functions capabilities have been incorporated in the PLCs. Ladder diagrams are best suited for discrete control and interlocking.

8.2.2 Structured text (ST)

Structured text is very close to traditional computer programming, similar to high-level programming languages like C or Pascal language. It utilizes standard programming constructs. Structured text is used for high-level functions such as complex logic or supervisory control. Structured text can be used to perform any form of control.

Following are some standard constructs used in structured text:

- IF-THEN-ELSE condition statement
  IF <condition-1> = True THEN <action-1>
        ELSE IF <condition-2> = True THEN <action-2>
        ELSE <action-3>
        END_IF
- WHILE-DO condition statement
  WHILE <condition> = True DO <action>
  END_WHILE
- FOR-TO-DO condition statement
  FOR X = a TO b BY c DO <action>
  END_FOR
- CASE statement
  CASE variable OF
        X: <action-1>
        Y: <action-2>
        Z: <action-3>
        ELSE <action-4>
  END_CASE

8.2.3   Instruction list (IL)

Instruction list programming is similar to assembly language programming. Instruction list consists of pneumonic code-based instructions and operands. Instruction lists can be used for programming any type of control. Instruction list programming is very cryptic and difficult to interpret and understand for the person who has not written the program. However, IL programming has found widespread use in industrial automation applications.

8.2.4   Function block diagram (FBD)

A function block (FB) is a graphical representation of a control algorithm. The FB is implemented with some internal codes. In some cases, that may be of other IEC 1131-3 languages, but normally it is the actual FB from the library, which is provided with the control system. Placing a given FB in FBD configures FB. A FB can have inputs and outputs that are linked either to variables in the control system, constant values or another FB. An example of a FBD is shown in Figure 8.1.

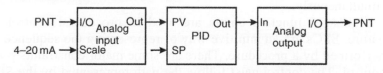

Figure 8.1
Example – function block diagram

Operations are made up of phases that may run concurrently and have either complex or product-specific inter-relationships or both. A list format can be used for simple serial cases, and SFC that allow representation of a range of sequential relationships are useful in more intricate cases. Unit procedures are made up of operations that are usually executed sequentially. Therefore, a list format or sequential function chart may be used to represent the operations in a unit procedure.

8.2.5   List format

A list is the simplest way of representing a linear sequence. A list has the advantage of being easily visualized and precise. However, a list can be useful only in simple cases as it is difficult to depict parallel and complex sequences. Though difficulties with the list format to represent complex process, a list format is adequate when the recipe procedure is simple or is simplified with engineering effort. The list format for an operation specifies what phases are executed and in which sequence. The list of phases is displayed in tabular form where phases are listed together with associated key information and parameters.

8.2.6   Sequential function chart (SFC)

More complex recipe sequences require representation that can clearly depict a variety of ordering logic. One of the practical methods is function chart is defined in the IEC 848 standard.

This is commonly known as SFC and has following advantages:

- Flexibility
- Broadly understood
- Well documented.

The SFCs are used to program in a defined order sequentially. SFCs consist of steps and transitions. Each step points to another SFC or another form of control – ladder diagram (LD), instruction list (IL), structured text (ST) or function block diagram (FBD). Transition causes the control execution to move from one step to another. A typical sequential function chart is shown in Figure 8.2.

In the sequential function chart, boxes represent steps and short horizontal lines represent transitions. The sequential function chart is processed from top to down and from left to right. Multiple, single horizontal lines at the same level correspond to conditional paths. These conditions are processed from left to right. The first transition to be true causes the control after that transition to be performed. All transitions and paths to the right side of the first true transition are ignored. Double horizontal lines in SFC represent parallel processing. All paths under the horizontal lines are performed simultaneously.

Sequential function charts are used in procedural control due to their sequential nature. SFCs are an intuitive way of representing the sequence of actions that are to be performed by a procedure. There are some major constraints placed on execution of the control. The control must follow the path represented by the SFC. There is no jumping of control if an exception condition necessitates a move from one part of the chart to another.

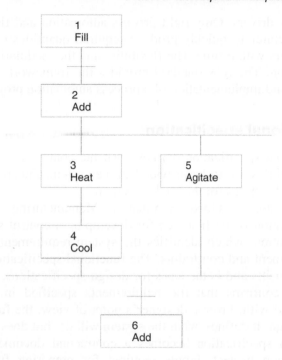

Figure 8.2
Example – sequential function chart

8.2.7 Guidelines for use of SFC for phase programming

Following are certain guidelines on usage of SFC for phase programming:

- At times parallel steps need to be performed within a state that involves branching and decision-making. Within a single state it is recommended to use not more than two parallel branches.
- Nesting of SFC levels is not recommended within a state.
- SFCs do not support interrupt handling.
- If sub-programs are launched by the SFC, it should have only single entry and exit point.
- Any divergence should be followed by a convergence.
- Do not have timers in transitions.
- The transitions on a divergent path must be mutually exclusive so that only one branch is executed.
- Limit a single step to a single function.
- Limit transitions to a single expression.

8.3 Practical methods of functional specification

Normally, the process automation requirements are documented in functional specification which follows from the process design. Process engineers, who define the process control sequences and critical control parameters for each unit, are sometimes not familiar with S88 standard concepts as the control or automation engineers. This results into the functional specification effort for the first time that the process requirements are translated in terms of S88 standard batch process control requirements. Further, the requirements for flexibility and quick response to market needs have become important business drivers. Only right process automation and flexible batch control can enable a manufacturer to quickly produce required quantities of the products and deliver to the customers within time. The flexibility and the modularity can be incorporated only in the beginning. The S88 standard provides the framework for addressing these needs in the design and implementation of a process automation project.

8.3.1   Functional specification

Functional specification defines what the control system should do and what functions and facilities are to be provided in the system. Function specification provides a list of design objectives for the control system.

According to Good Automation Manufacturing Practice (GAMP), the function specification is the first step for developing a control system based on user requirement specification, which identifies the system requirements with regard to data, interfaces, environment and constraints. The functional specification defines the process automation requirements and is the basis for design specifications. Acceptance testing of the control system confirms that the requirements specified in the functional specification are complied with. From a designer's point of view, the functional specification is the basis for design. It defines, what the system will do, but does not specify how to do it. Usually, function specification becomes a contractual document that defines the scope of an automation project. Inputs required for preparing functional specification include – process description, process flow diagrams, instrument list, and piping & instrumentation diagrams (P&IDs). Any shortcomings in developing comprehensive function specification can result in project delays and unsafe process automation implementation. The S88 standard concepts help in ensuring the creation of accurate and practical functional specification for a process automation project.

The person who prepares the functional specification needs to decide the boundary between functional specification documents. To have a single functional specification document for the entire manufacturing facility, or to have small functional specifications/documents for each individual module. One single functional specification for the entire manufacturing facility has the advantage of providing one source of reference. It provides a big picture of how the facility works, and it is convenient to maintain and get approvals. On the other hand, small function specifications/documents have advantage of easy manageability. These advantages and disadvantages for functional specification documents must be evaluated for each automation project specifically. As for a small automation project, it may make sense to use a single functional specification document, but for large automation projects, a single functional specification document may become unmanageable. Normally, for a large project it is better to create functional specification documents for each area. This allows description of classes within a single document as classes often fall within a process area.

Once the functional specification is prepared, it is time to involve all stakeholders. All the persons involved in development of the functional specifications should review the document and finally approve the document. All stakeholders should agree on the requirements before the start of design and implementation.

8.4   Defining equipment procedures

As discussed in Chapter 7, a procedure is the strategy for performing work to make a product. There may be many alternatives for mapping of procedural elements and equipment. There may be units, equipment modules and control modules within a process cell. An equipment phase may exist within a unit or an equipment module. An equipment phase may act on equipment module(s) or control module(s).

8.4.1   Treat modules as objects

Equipment modules and control modules should be treated as objects in a phase. A phase should not directly act on an output. A phase should send a message to the control module requesting it to take an action.

Let us consider an example where a phase acts on a valve treating it as an object, as shown in Figure 8.3. The phase sends an 'open' message to the valve control module. The valve control module then takes appropriate action to open the valve and verify the status of the valve position. The phase then verifies that the valve is open with the open feedback from the valve control module. The valve control module takes care of the valve status and gives the valve open position feedback.

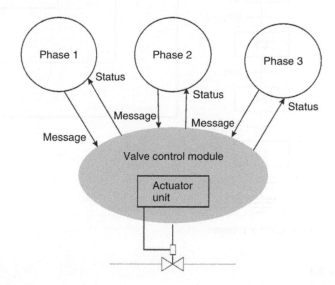

Figure 8.3
Phase to control module communication

8.5   Phase logic programming

Phase programming should be done on the basis of clearly defined set of templates for each phase. Then, for each phase the actual task to be performed during the phase is plugged into the template. Ladder logic diagram, ST and SFC are used for phase programming.

8.5.1   Phase logic – sequential and parallel programming

For phase programming using sequential function charts, there are two methods – sequential or state machine.

Each method has pros and cons as described below:

- *Sequential programming*: Sequential phase programming is more intuitive and easy to trouble shoot. The sequential function chart shows the order in which the steps are executed, as shown in Figure 8.4(a). It is difficult to make changes to SFCs. In order to change the order of the steps, the SFC must be redrawn. SFCs also have problems with recovery from an exception. Each time SFC is started, it must start from the first step and progress through connected steps and transitions. For a restart from a step other than first step after an exception, it is a must to show a transition from the starting point and every step in the SFC, as illustrated in Figure 8.4(b).

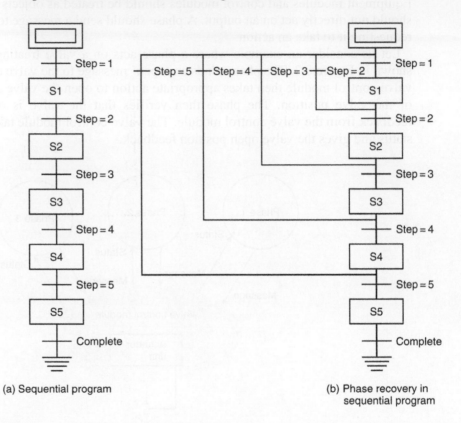

(a) Sequential program

(b) Phase recovery in sequential program

Figure 8.4
Sequential programming

- *State-machine programming*:   State-machine programming can be implemented using sequential function chart or structured text. An example of state-machine program implemented using SFC is shown in Figure 8.5.

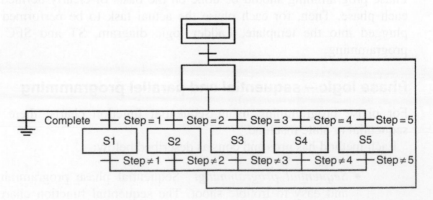

Figure 8.5
State-machine programming

The structured text program for the same is as follows:

- WHILE (NOT complete) DO
  ```
        IF Step = 1 THEN S1
        ELSE IF Step = 2 THEN S2
        ELSE IF Step = 3 THEN S3
        ELSE IF Step = 4 THEN S4
        ELSE IF Step = 5 THEN S5
        END_IF
  END_WHILE
  ```

  In a state-machine program, the phase control can move from any step to any other step based on transitions. In state-machine programming, new steps can be easily added and existing steps can also be modified or re-sequenced easily. State-machine-based programming allows recovery from an exception without any changes. However, state-machine-based programming is less intuitive and does not show the sequential nature of the steps.

- *Single-step*:   Single-step mode is an aid that helps during troubleshooting and testing. When a phase is in a single-step mode, the transitions are not processed further until the operator gives a resume command. On resume command from the operator, the phase processes the transitions and moves to the next step and again waits for next resume command from the operator. This helps in testing and verifying the control actions executed by each step within a phase. However, due care must be taken to ensure that a phase should not get into a single-step mode during normal process operation. A single step can be implemented as follows:

  ```
  IF Single-step THEN Paused_Flag = TRUE
  END_IF
  ```

  The paused flag is used for disabling the processing of transitions within the phase.

- *Pause*:   Using pause mechanism, the operator can halt the execution of a phase at a safe point. When a phase is in the paused state, the transitions are not processed until the operator gives a resume command. After a resume command, the phase continues normal operation. A pause can be implemented as follows:

  ```
  IF Pause THEN Paused_Flag = TRUE
  END_IF
  ```

- *Resume*:   Resume is a mechanism required for resuming the normal operation after a single-step or a pause. On receiving resume command, the phase moves to the next step in case of pause after single-step and continues normal operation in case of pause due to pause mechanism. A resume can be implemented as follows:

  ```
  IF Resume THEN Pause_Flag = Flase; Paused_Flag = False
  END_IF
  ```

- *Phase state model*:   For designing individual phases or a phase template, first a state model for the phase is determined. Most of the batch management systems specify a phase model. A typical phase model is illustrated in Figure 8.6. For the phase state model, various states, running, restarting, aborting, stopping, etc., must be programmed. The running logic is the normal phase operation. The holding logic is the exception handling logic and the restarting logic handles recovery from an exception. Aborting logic includes the logic to abort execution of the procedure. In the state model, abort is used to abort execution of the batch and hold provides the exception handling logic. The stopping logic has the logic to stop the phase execution when it is externally terminated.

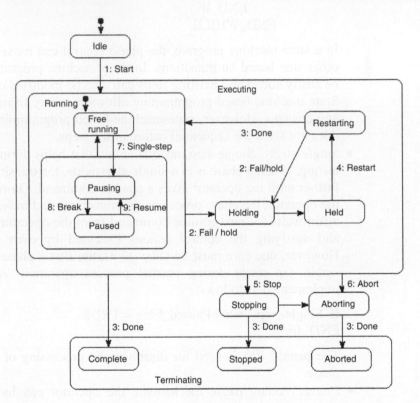

Figure 8.6
Phase-state model

- *Synchronization of phase*: In a batch management system, capability to coordinate the execution of phases is a must. The capability of coordination of execution of phases should be used whenever possible. Capability of synchronizing phase management in a batch management system reduces the amount of peer-to-peer communication between the PLCs. The amount of work needed to change or add phases is also minimized.
- *Allocation and arbitration*: The allocation and arbitration functions should be performed by higher-level system, as this will minimize programming changes in future due to changes in downstream or upstream processes, or development of new products.

8.6   Phase logic interface

The phase logic interface is used for tying procedural and physical entities as illustrated in Figure 8.7. The phase logic interface connects a batch server to the phase logic in the PLC. Phase logic is programmed in the PLC and provides interfaces through a defined set of PLC registers and bits. The batch server interface processes server command, reports phase status and makes requests of the server. The phase logic interface defines the phase logic's operating state and receives status information from the phase like completion, failures, etc.

Control recipe procedure

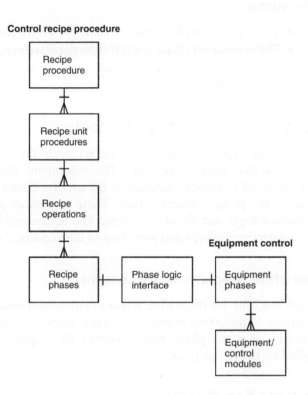

Figure 8.7
Phase logic interface

8.6.1 Equipment control information exchange

The ISA S88 Part 2 defines data structures and communication protocols that can be used for communications within and between batch control systems. Communication requirements between the recipe phase execution and equipment phase execution are defined by the specification briefly described in the following section. However, it is important to note that this standard may not be applicable to all batch control applications.

8.6.2 Recipe phase and equipment phase interface

Generally, equipment phase logic has control over the events that are performed. The recipe phase starts the equipment phase, but after starting the equipment, phase logic takes over the control of most of the events. The communication interface consists of common state machine enforcement between the recipe and equipment phases, a set of services that supports requests from equipment phase to the recipe phase and a set of services that supports commands to the equipment phase. The recipe phase has various service supports such as arbitration that it can provide to the equipment phase. These services allow the phase logic programmers to write completely independent and stand-alone equipment phases. This, to a great extent, simplifies the efforts required in phase logic programming. It also allows the equipment phases to be residents of multiple process control devices, even those that do not support peer-to-peer communications.

The recipe phase expects from the equipment phases to conform to the following specific criteria:

- The equipment phase must follow the state transition diagram
- The equipment phase must follow the interface protocol
- The equipment phase should use specific requests to access the services provided by the recipe phase.

8.6.3   Functions of equipment phase logic

An equipment phase designed to interact with the recipe phase has some specific requirements that need to be met. The equipment phase logic should be written to implement set of sequences defined by the user for direct control of process equipment to accomplish the process-oriented task. These sequences perform normal logic as well as the abnormal logic and shutdown logic. The equipment-phase logic should also include logic for detecting failures and reporting of the sequence steps.

8.6.4   Phase-state machine enforcement

The state machine for the recipe phase to equipment-phase interface defines a model for the behavior of these two objects. The main purpose of the phase-state transition diagram is to ensure that the phase logic exhibits the required behavior characteristics to be controlled by the recipe phase.

8.6.5   State transition diagram

The state transition diagram as shown in Figure 8.6 illustrates the active and quiescent states that are supported by the logic and the paths between these states. The state transition diagram also illustrates the states and transitions of the communication protocol between the recipe phase and the equipment phase. The phase logic must adhere to the rules depicted in the state transition diagram. Only valid state transitions as depicted in Figure 8.6 may be utilized. Though the configuration of the phase logic will vary from site-to-site, the logical constructs should always conform to the state transition logic.

8.6.6   Events causing state changes

In the state transition diagram, the transitions between states are caused only due to following three types of events:

Commands

The first type of state transition is commanded state transition. The phase logic is expected to respond to the legal commands depending on its present state. For example, when the phase logic is in idle state, and a start command is given, it will cause phase logic to transit from idle state to the running state. However, if the phase logic is in idle state and a stop command is given, this is treated as an invalid command and reported as a communication error. Legal commands for each phase logic state are illustrated in the state transition diagram. Following is the brief description of various commands:

- *Start*: Causes a transition from idle state to running.
- *Stop*: Causes a transition from running, holding, held, pausing, paused or restarting state to stopping.
- *Hold*: Causes a transition from running, paused or restarting state to holding.
- *Restart*: Causes transition from held state to restarting.
- *Abort*: Causes a transition from running, holding, held, pausing, paused, restarting, stopping or stopped to aborting.
- *Pause*: Causes a transition from running to pausing.
- *Reset*: Causes a transition from aborted, complete or stopped to idle.
- *Resume*: Causes a transition from paused to running.

Completion of equipment phase logic

During execution of logic, state changes occur as the result of completion of one of the following states:

- Running
- Holding
- Restarting
- Aborting
- Stopping.

On completion of sequencing, each of this state is expected to transit to another state. The end states on completion of sequencing are:

- Completed
- Held
- Aborted
- Stopped.

Response to a failure

Response to a failure may cause a state transition. On failure detection during the running and restarting states, the phase logic is permitted to transit to the holding state.

8.6.7   Active states

Running

Running is the normal active state of the equipment phase. The phase logic processes its normal execution path. Within the running state, the phase logic may operate in either a free-running state or a breakpoint-step state. For breakpoint-step state, the phase logic may be defined to support paused and pausing states, as shown in Figure 8.6. Here, the paused state is quiescent state. Figure 8.8 illustrates the running state changes depending on three events. For example, let us consider the example of making raw material slurry batch for fiber–cement sheets manufacturing. For cement dozing phase, in Mixer-2, during the running state the phase logic will be sequencing the valves, cement weigh feeder, monitoring cement Feedrate or Mixer-2 load cells.

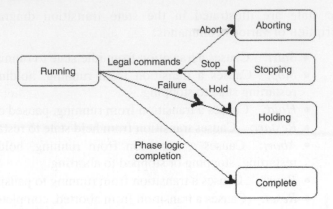

Figure 8.8
Running state changes

Holding

During holding state, the phase logic performs necessary logic to transit phase devices to a known state from which the batch execution may be resumed. The actual sequencing may vary depending on the state of the phase's devices and logic when transition to the holding state had occurred. Transition from holding state to other states, depending on occurrences of various events, is illustrated in Figure 8.9.

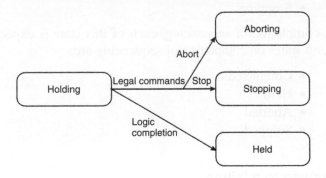

Figure 8.9
Holding state changes

Restarting

The phase logic performs necessary logic to transition from a held state back to normal execution path. However, the actual sequencing may vary depending on how far the phase logic had progressed before transition to the held state. Transitions from restarting state to other states, depending on occurrences of various events, is illustrated in Figure 8.10.

Aborting

During this state the phase logic performs necessary logic for an abnormal transition from the current state to a known state in which phase logic processing has completed. However, the actual sequencing to execute the controlled normal transition to the aborted state may vary depending on the state of the phase's devices and logic when the transition to aborting state occurred. The aborting logic is expected to perform a rapid transition to the aborted state. It differs from the stopping logic, which is expected to perform an orderly shutdown.

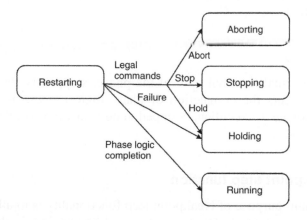

Figure 8.10
Restarting state changes

Stopping

The phase logic performs necessary logic for a controlled transition from the current state to a known state in which the phase logic processing has completed. However, the actual sequencing to execute the controlled normal transition to the stopped state may vary depending on the state of the phase's devices and logic when the transition to the stopping state occurred.

8.6.8   Quiescent states

Idle

In idle state, the equipment phase is dormant and inactive. Idle state is considered a safe state and expected to be the default initial state of a phase. Idle state transition to running state takes on start legal command, provided no failures are detected or present during the idle state.

Complete

Complete is the normal terminal state of the phase. Complete is also considered to be a safe state. It represents completion of the phase logic processing and indicates that the phase is ready to be reset for the next execution.

Held

Held state represents that the phase logic has transitioned to the held state in middle of the phase logic execution from where the batch execution will be resumed again. It is important to note that the definition or terminology of the held state does not infer that nothing is happening. The definition of held state may require functions such as agitation, temperature control, etc.

Aborted

This is an abnormal terminal state of phase. Aborted is considered to be a safe state. It represents that the phase logic processing has been completed and the phase is ready to be reset for its next execution. Aborted state transition to idle state occurs on RESET legal command.

Stopped

An abnormal terminal state of the phase, stopped is also considered to be a safe state. It represents the completion of phase logic processing, and the phase is ready to be reset for next execution. Stopped state transition to idle state occurs on RESET legal command.

Breakpoint step function

The main purpose of breakpoint step functionality is to allow the user to force equipment phase execution to stop at programmed breakpoints in the equipment phase sequencing and wait for operator confirmation before proceeding further. The breakpoint step functionality within the equipment phase API was defined to allow users to take advantage of programmed optional breakpoints in the phase's sequencing logic. However, the actual programming of breakpoints into the phase's sequencing logic depends on the individual project application team.

When the breakpoint step bit is on, it indicates that the phase is in breakpoint step operation and the equipment phase sequencing will pause at every programmed breakpoint in the equipment phase. At a sequence breakpoint, the equipment phase halts sequencing and the paused flag is set on to indicate that the equipment phase execution has been suspended. The equipment phase is resumed when the user gives a resume command. The phase execution resumes until the next breakpoint step is reached or the user gives a breakpoint step command to switch the equipment phase completely out of breakpoint step mode and resume the normal execution. The sequence breakpoints may or may not be safe points in the equipment phase. In most cases, the breakpoints are not process-safe points as they are intended for startup and commissioning checkout of the equipment phases and not for normal operation. The breakpoint step functionality is a useful tool during the startup and commissioning phase of a project. The breakpoint step functionality is less used as an optional part of normal batch execution.

8.6.9   Exception handling

An exception is an event that occurs outside the normal or desired behavior. An exception can occur at any level in the control activity model. It may be a part of basic coordination or procedural control. An exception may affect the modes and states of equipment entities and of procedural elements. For exception handling, an exception event must be detected, evaluated and a response must be generated.

The ISA S88 standard defines exception handling as functions that deal with the plant or process contingencies and other events, which occur outside the normal or desired behavior of the batch control system. The S88 standard provides models such as procedural state matrix and the terminology for addressing exception handling. However, the person who specifies the control system requirements should specify specific details of identifying the exceptional conditions and determining the appropriate reaction.

There are certain considerations that must be made while specifying exceptional handling requirements. The considerations for exception handling requirements are:

- The exceptional handling must be considered as an integral part of the automation requirements. Significant part of specification development is spent on this activity. If exceptional handling is not addressed adequately, then the plant equipment, and product integrity and consistency may be doubtful.
- An exception and its corresponding reaction must be considered as a pair. The reaction to an exception should fit the severity of the exception. For example, in one case, it may be sufficient to send an alarm to the operator. Other critical exceptions may require the batch to be aborted. Further, the reaction may change relative to the current step of the process. For example, a high-temperature alarm for an ID fan new bearing may not require the same reaction as when the bearing temperature is stable.
- In the phase logic, it is not only important to consider how to react to an exception, but also how to recover from the exception. In case of some exceptions, recovery may be simple, such as taking an alternate action in the running logic. However, in case of critical exceptions, recovery may be complicated. The most complicated recovery is from exceptions that require the phase to go to a safe state, such as holding. Such recovery can be labor-intensive, as one must logically coordinate the actions of the running, holding and restarting states.

Exception handling is best done with states. What states are needed in exception handling must be defined by the user. The user must also define the conditions that cause state transitions.

9

Operator and supervisor interface

9.1   Introduction

A properly designed batch control system should make it easy to describe the sequence of operations and the checks that must take place. It must provide displays and interfaces, which the operator can use to monitor and direct the process activity. An advanced batch control system should provide displays which are oriented toward the total process rather than towards individual parameters. User interface is the mechanism by which the operators and supervisors interact with the systems. The interaction may be active such as controlling the process or may be passive such as simply monitoring the system. The user interface consists of devices that help operators and supervisor to monitor, control and supervise a process and annunciate alarms. The user interface provides information in the form of text, pictures, audible alarms and color changes, etc. Most systems provide tools to the users for building user interfaces. While developing the user interface, it is important to keep in mind the profile of its user. The user profile should include age, cultural background, education, goals, personality, physical abilities and motivation. An operator may need to know the status of the process and provide inputs to control the process, whereas a supervisor or a manager may need an overview of the process with some performance indicators. A maintenance technician may need to have access to a device and its parameters to troubleshoot the device. A system administrator requires system status for fault diagnosis and troubleshooting. The user interface standard may focus only on the custom displays through which the process interaction takes place.

Windows system is the most popular technology for user interface since it has the following advantages:

- Different types of information can be displayed simultaneously
- Multiple tasks are available through pull-down menus
- Systems are more user-friendly.

9.2   Display screens for batch management

9.2.1   Operator interface

The operator interface provides the operator with a user-friendly interface through which she/he can get information about the status of the plant and interact with the plant. In manual mode, the operator can give commands such as open/close valve, start/stop pumps, set points to PID controllers, etc. The operator interface consists of graphic displays that give real-time status of plant equipment, drives and process parameters. Operator interface also provides the operator historic trend displays, alarm annunciation, status of batch execution, etc.

An operator interface for batch management system includes:

- *Batch list view*: A list of all batches/control recipes and their current states.
- *Sequential Flow Chart (SFC) view*: A graphical view of a single-control recipe.
- *Tabular view*: Provides a view of a single-control recipe in spreadsheet form.
- *Prompts for operator inputs*: A list of all prompts from all control recipes that require operator input.
- *Phase control display*: An interactive display that allows manual execution of individual equipment phases.
- *Phase summary*: A global view of all the equipment phases configured within the area model and their current states.
- *Alarm summary*: A display of alarms, error messages and failure messages for current control recipes. Area-wise and unit-wise alarms displays and display of overall alarms in the plant.
- *Events journal view*: A real-time record of batch that provides details of events related to a particular batch. Area-wise and unit-wise event journal displays.
- *Resources allocation view*: A view of current resource allocations. Resource allocation can be arbitrated and manipulated by the operator.
- *Configuration function*: Provides the means for configuring the custom graphics, displays, group trends and other displays as per the operator's or supervisor's needs.
- *Help function*: Provides online help to the operator.
- *Security function*: Security login dialog boxes, to control access levels.

The above-listed display screens for recipes, monitoring sequence operations, maintenance and troubleshooting are the standard display screens provided by most of the batch management software packages available in the market. However, each software package may have additional displays with slight variations. Further, the software packages provide facilities for creating custom displays.

Various display screens available in batch management systems shall be illustrated during practical session using the software packages.

9.3 Guidelines for building user interfaces

The human–machine interface (HMI) or man–machine interface (MMI) is all in the head. In today's world, humans interact more with computer-based technology than with chisels, cutters, drills, hammers, etc. Unlike tools, the visible shape and the controls of a computer do not necessarily communicate its purpose. The task of a HMI is to make the function of a technology obvious. As a well-designed hammer fits the user's hand and makes a physical task easier, so too the well-designed HMI fits the user's mental map of the task he wishes to carry out. The effectiveness of the HMI can predict the intended users' acceptance of the entire solution. As far as plant operators or system users are concerned, the HMI is the product. The operator's and system user's experience with the MMI is more important than the architecture of the internal workings of the system.

9.3.1   Design with the user in mind

The HMI in the industrial environment is the channel of communication between operators and the plant that ideally provides them with process parameter and information necessary to keep the plant running at maximum efficiency. To design a successful HMI, one must understand the way in which humans think conceptually and understand how the operators or system users process this information physically. The physical and conceptual thoughts are closely related when working on any type of interface. Mayer and Shneiderman developed the syntactic/semantic model of user behavior to describe programming. As per the syntactic/semantic model, operators and system users organize computer knowledge either syntactically or semantically.

Syntactic knowledge

Syntactic knowledge is short term and based on device-dependent details of a system. It is often learned by rote and frequently void of relationships to a system. An example of syntactic knowledge is the popular use of meta keys or combination keys such as 'ctrl-p' or 'ctrl-shift-down arrow'.

Semantic knowledge

Semantic knowledge proliferates through concepts, relationships and analogies. Semantic knowledge is task-based and is difficult to forget after it is learned. It is often conveyed using pictures. An example of semantic knowledge is how certain graphical user interfaces use a picture of a file cabinet for their file managers. The relationship of computer files to the way file folders function in an office is not abstract. System users will have an idea of what the program does just by its pictorial representation. Using the concept of semantics in designing a HMI can significantly reduce learning time, anxiety and stress.

9.3.2   Lay out the plant on brain

Mental models are mental pictures of a plant or operation that help operators understand the complexities of their plant. Understanding mental models is fundamental to the success of a HMI because the designer is in essence recreating a mental model. The problem with many mental models is that they are incomplete, inconsistent, oversimplified, unstable in time and rife with superstition. Any group of operators will invariably describe its plant's workings differently. Construction of an operator interface is a construction of the operator's mental model, and construction of a HMI necessitates understanding all the operators' models. It is easier for the designer to fit the mental model of the operator than it is for an operator to understand the designer's model.

9.3.3   ISO 9241 standard

The International Organization for Standardization's (ISO) Standard 9241 defines three components of quality that are applicable for designing HMI with effectiveness, efficiency and satisfaction. As the designer cannot pull out a meter and calculate these intrinsic product qualities, the crux of the ISO 9241 component would seem to be that the unifying principle of these design techniques is keeping the user at the center of the design process. Mental models translated into renderings of ground floor trends and floor plan.

9.3.4   Integrating systems

One of the key reasons for integration of distributed control systems with PLCs and computers is to obtain a superior HMI. PLCs do not have an embedded HMI except push-button switches and indicator lights. The distributed control systems have excellent color graphics displays in the form of an operator's workstation. To take advantage of this robust and user-friendly interface, the need for integrating the PLC and DCS was felt, as PLC and DCS need to share information. The PLC provides the status of the controlled devices to the DCS, and the DCS provides the PLC with control signals, which will start or stop at particular motors or group of motors or open and close valves. This integration visualizes at the HMI. The HMI informs the operator when a requested action is inhibited and advises the operator what is preventing the action from occurring. For example, if a belt conveyor motor start is requested but is inhibited because a belt sway switch is actuated, rather than just not starting the belt conveyor motor, a well-integrated system will describe the nature of the problem to the operator by changing the color from red (stopped) to yellow (drive not ready for start). The use of color and shape coding to improve the operator's ability to perceive relevant information is a highly debated topic. Color in the human brain processes in parallel, while shapes process serially. In a screen that is highly congested, color can help highlight special groups of information. Grouping is perhaps one of the most efficient uses of color for display purposes. Because humans have trouble distinguishing the meaning of more than four colors simultaneously, colors should be used sparingly and consistently to add to the understanding of the desired message.

There are several principles that guide the user interfaces design standardization. Some of the common guidelines are:

- *Consistency*:   A consistent format must be used in the menu structure, display elements, graphics and interaction with the functions of the system mode and state changes, etc. In other words, for changing a parameter one should not provide different ways of doing it at different places in the system.
- *Feedback*:   To ensure proper communication, provide feedback.
- *Verification*:   Any critical entries to the system should be verified. To prevent any accidental entries, request should be confirmed by two keystrokes. Example – emergency stop or change of mode from auto to manual to vice-versa.
- *Organization*:   Organization of elements on the interface is important and applies to screen layout and functions. The screen layout should be such that the users can easily learn where to look for information. The functions should be organized in an intuitive manner.
- *Choice of elements*:   The selection of elements should be carefully done. Say, if a combination of touch screen, and point and click device are used, the touch screen needs to have verification for data entry. Complex structures should be avoided for providing informational displays.

9.4   Consideration of human and ergonomic factors

The design of an effective MMI for a process plant is a complex task and involves a variety of human-related issues and ergonomic factors. For safety and efficient operation of the plant it is important that the operator's functions are designed such that he/she can perform effectively and at the same time suitable tools and techniques are provided to perform these functions. An alarm system is one of such support tools. It must be noted that however good and intelligent an alarm system is, an operator cannot run the plant safely and efficiently if his functions and tasks are beyond his/her capabilities, and neither the plant control system nor the MMI is sufficient to provide the required support. An alarm system should always be designed considering the overall functions of the operator and support tools and systems.

Ergonomics is a study-measurement-organization of work concerned with making purposeful human activities more effective. The focus of the study is the person or the operator interacting with the engineered environment. This person or operator has some limitations which the designer should take account of while designing the system. Complexities arise from the nature of human beings and the variety of the designed situations that are to be considered. The latter varies from simple ones like chairs, handles of tools, lighting, etc. to the more complex, like process control rooms in industrial plans, artificial life support systems for space, etc. The people studied can be pilots of spacecraft, racing car drivers, locomotive drivers, call-center operators, control room operators in process plants and others. At times, the relevant characteristic of the study is essentially biological and unchanging, except with age. Sometimes it varies with sex and race, e.g. body dimensions. Sometimes it even varies with the degree of cultural, economic and social development, e.g. acceptable working hours, or working in night shifts.

Ergonomics covers energy consumption studies but excludes the design of diets to provide the energy. Physical hazards are part of ergonomics but the chemical hazards are not. Selection of an operator does not fall in the scope of ergonomics but training techniques do. The design for safety and comfort is part of ergonomics but design of social satisfaction, quality of work life, industrial democracy, etc. are not. A successful ergonomic study is the one that increases the knowledge or productivity.

Anatomy, physiology and psychology contribute to ergonomics. Anatomy provides a conceptual background of anthropometry and biomechanics. Anthropometry is the measurement of man that provides the dimensional data needed for positioning of controls and the workspace dimensions. Biomechanics deals with the application of forces by the human body. Physiology provides details of work physiology and environment physiology in ergonomic studies. Work physiology mainly deals with the human process of energy production. Environmental physiology provides measures of stress and reasonable standards for environmental parameters like light, climate, noise, vibrations, radiation, etc. In each case there is no physical measure, which indicates the effect on the person.

Psychology also contributes in a number of ways to ergonomics. The main contribution is the human performance theory, which is based on an information model of the human operator. Other contributions include learning theory and skill theory in relation to work design and system design. Almost every psychological theory tells us something about human errors and why people make mistakes. The major part of human work is based on information processing and decision-making regarding people. Automatic machines provide great capabilities in terms of power and precision but it is the operator who selects and controls the particular operations based on information received from the general environment and artificial information displays. The operator through his eyes and ears receives information about the state of the equipment/machine. A thorough understanding of audio and visual mechanisms is necessary in order to design efficient presentation of information and alarms. Once the data/information is received through the eyes and ears, it must be interpreted and acted upon. The human performance specialists who provide the relevant knowledge on information processing capacity, memory, and attention and fatigue effects study such processes.

In the case of environmental variables there is an optimal range of information flow within which effective performance is maintained. Too little flow and we get problems like a lack of vigilance. Too much and we get problems like stress and overload. Ergonomics primarily deals with fitting the job to the worker. The best match of worker and job essentially requires a two-way approach and some consideration of adapting people to the jobs by giving them training. The key to successful training is to provide knowledge of results that under suitable conditions will not only orient the operator in the learning process but also have considerable incentive effects. Breaking down tasks into optimal learning units with associated knowledge of performance is an important contribution from psychology not only in training but also in maintenance and improvement of performance by skilled operators. Any work involves interaction with other people in the physical world so it requires some appreciation, interpersonal communication skills, moral, leadership, group behavior, teamwork, organizational behavior, etc.

While designing an effective alarm system, ergonomics can be used in various aspects of the design of work, systems, workspace, environment, MMI and work situations (see Figure 9.1).

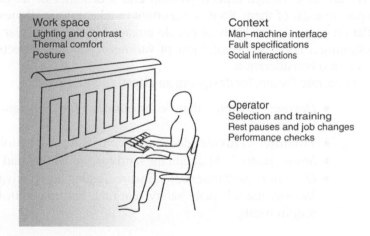

Work space
Lighting and contrast
Thermal comfort
Posture

Context
Man–machine interface
Fault specifications
Social interactions

Operator
Selection and training
Rest pauses and job changes
Performance checks

Figure 9.1
Various ergonomics aspects in the design of a man–machine interface

System design incorporates the principle of man as an integral part of a total man–machine system. The primary design problem is to allocate functions between man and machine or more generally between men, machines and procedures. The main reason for the success of a man–machine system is that each has characteristics that compliment each other. Man is intelligent, versatile and adaptive whereas the machines are powerful, fast and tireless. The man–machine system is very effective if care is taken to ensure that the main components are allocated functions matching their advantages/strengths and limitations/weaknesses. Good designers have always done the allocation of functions intuitively, but the system ergonomics ensures that it is done systematically. Doing this involves a set of knowledge and techniques that is new and not derived from old human sciences.

Workspace design on the other hand ensures that the physical surroundings fit the characteristics of the human body so that the work can be done without excessive efforts within the range of healthy postures, standing, sitting or exerting forces. Workspace design involves seat design, console or control desk design and positioning of displays, controls and materials.

Environmental design ensures that the lighting, heating, ventilation, vibrations, noise, etc. are appropriate to the requirements of the human operator.

MMI design focuses on the exchange of information between the man and the machine or environment. Exchange of information is bi-directional; displays present information to the operator and the machine receives control commands and information from the man. Display design incorporates issues like optimal design of scales, pointers, sizes/fonts/color of letters and numbers, positioning and grouping of machines/equipment, etc. Presentation of information is an elaborate topic concerned not just with the design of symbols but also with the rules for making their combinations and meanings in relation to different tasks. Design of information presentation on CRT/video displays is the main topic of interest. Control design has anatomical aspects concerned with aspects like sizes, shapes, positions, forces and psychological aspects such as discrimination and identification. As usual, the control design focuses on the attributes of the operator, some of which are innate and some required.

Work situation design deals with the broader issues like hours of work, rest pauses/breaks, problems related to shift work, interpersonal and organizational aspects of the work.

As all these design aspects overlap and a designer or an ergonomist may not be an expert in each of them. So it is important to take an overall view of key design aspects in relation to particular kinds of people engaged in the particular tasks. Like in the case of designing a HMI, an overall view of various ergonomic aspects of the operators must be taken into consideration.

Ergonomic factors for designing an effective MMI are:

- *Design*: System that compensates for inherent human and hardware limitations.
- *Training*:   Provide adequate training and work instruction to the operators.
- *Maintenance*:   Maintain the hardware reliability and performance.
- *Operator performance*: Ensure adequate motivation, prevent excessive fatigue, use adequate safety margins, and maintain basic speed and accuracy requirements.

10

Batch management issues

10.1 Introduction

Batch production management involves various activities like process control, unit supervision, process management, production planning and scheduling, production management, management of production information, recipe management, etc. To manage batch production successfully, it is important to ensure that all these activities and functions are planned systematically and implemented as per the planned schedule. The batch management system must also address issues related to safety system implementation.

10.2 Control activity model

As discussed above, batch production management involves many control activities and control functions. To manage batch production successfully it is important to ensure that all these control activities and functions are implemented successfully. The ISA S88 standard provides the control activity model as shown in Figure 10.1. It provides overall perspective on batch control and identifies major batch control activities and their inter-relationships. The control activity model is a functional model that helps in describing the functional requirements of a batch control succinctly. Though the standard does not design a solution nor does it provide implementation, it does help in the selection of a right batch control solution. Using the standard, the batch control functional requirements can be met by a solution implemented using any vendor application technology.

The control activities described in the control activity model are directly related to the real needs of batch manufacturing. Control activities from lower hierarchical level to the higher are briefly described as follows:

- *Process control*: Process control, as a control activity, encompasses procedural and basic control, including sequential, regulatory and discrete control in addition to data collection and displaying the data. Process control, as a control activity, describes the control functions that directly deal with equipment actions. Process control provides discrete and regulatory control for units, equipment modules and control modules. Process control involves the following control functions:
    - Execution of phases based on parameters received and commands
    - Collection of process data from sensors, derived values and events occurring during phase execution

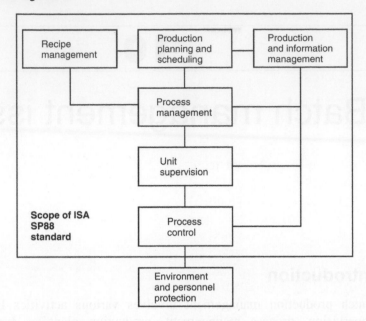

Figure 10.1
The control activity model

         – Execution of basic control that causes changes in the equipment and
           process states
         – Provide data/information to production information management
           system.

- *Unit supervision*: The unit supervision control activity involves coordination
  of process control activities, allocation of resources, supervision of execution
  of operations and phases. Unit supervision ties the recipe to equipment control
  via process control. Unit supervision involves the following control functions:

     – Acquisition of required units and execution of operations
     – Managing unit resources for both the acquired units and the other
        service-providing units
     – Collecting unit information and batch information and providing it
        to production information management system.

- *Process management*: The process management control activity involves
  control functions like creation of control recipes from master recipes, defining
  each batch as an entity, initiation and supervision of individual batches,
  coordination of unit activities, logging and report generation. The domain of
  process management is the process cell. Process management controls batches
  and resources within a process cell. The successful execution of a control
  recipe makes a batch, and process management is finished with the batch when
  the control recipe is complete. Process management involves the following
  control functions:

     – Managing batches by creating a control recipe from the master
        recipe based on equipment and scheduling information
     – Managing resources within a process cell by allocation, reservation,
        and arbitration of conflicts
     – Collecting and providing process cell and batch information to
        production information management system.

- *Production planning and scheduling*: The production planning and scheduling control activity creates the batch schedules required for process management to ensure production of batches within planned time schedule. Production planning and scheduling takes care of market demand, and ensures timely batch production and deliveries.

  A number of different types of plans and schedules are typically needed within an enterprise corresponding to various hierarchical levels in the physical model described in the ISA S88 standard. The S88 standard describes only the batch schedule, which meets the scheduling needs at the process cell level.

  It defines which batches are to be made, their order and the equipment to be used based on the specific resources and requirements of the process cell. The batch schedule is provided as input to the process management. During the actual manufacturing of a batch, real-time information is going in the opposite direction so that schedules can be updated within a short horizon. Scheduling is an important activity within a plant. However, finding a feasible schedule which reduces lead times and costs is often a complex task. Most of the methods and tools for production planning and scheduling focus mainly on the needs of discrete manufacturing industries. Real world problems in the process industries tend to be more complex, and general concepts for production planning and scheduling are hard to find, with some exceptions.

  Production planning and scheduling includes control functions such as development and revision of production schedules, and determining the raw material and equipment availability.

- *Production information management*: The production information management control activity takes care of the need to collect and store necessary information related to the batch production and to create and maintain batch histories. Production information management collects, stores and processes batch production related data/information and generates reports. This information can be used for batch analysis and reporting and also for other purposes. The static information related to the functional model can be found in the domain model, which organizes, and represents information related to the real world entities in a process plant.

- *Recipe management*: The recipe management control activity takes care of the need to have control functions that can create, store and maintain general, site and master recipes. The overall output of recipe management control activity is a master recipe that is made available to process management, which uses the master recipe to create a control recipe.

10.2.1   Extended control activity model

As discussed above, the ISA S88 Part 1 describes a control activity model. The model needs to be extended, as the scope of plant operation is wider than just batch control. As shown in Figure 10.2, one of the extensions is engineering activities such as maintenance, which are closely related to, but outside, batch control. Various maintenance functions to keep process equipment, instrumentation, knowledge and skills in good shape are essential for all other operational functions and creating an important foundation for safe and effective plant operation. Another important issue in plant operation is inter-departmental coordination, such as production vs support service. Aspects of these issues are illustrated by another extension of the control activity model by including organization information. There are other crucial and critical functions to the plant operation that are not included in this functional model, such as quality control, customer service support, inventory control, process and product development, regulatory reporting and process validation. These functions are defined outside the boundaries of the functional model and thereby outside the scope of batch plant operation.

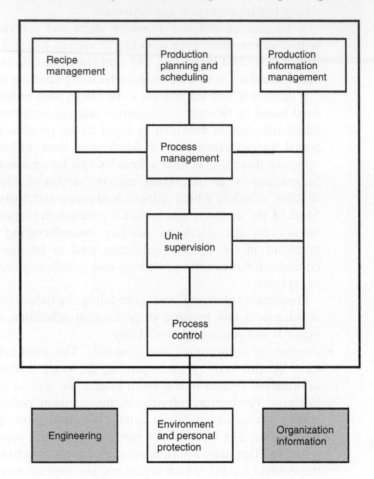

Figure 10.2
The extended control activity model

10.3 Practical problems in batch management

Plant operation in a modern batch plant is a complex task as it includes a wide range of activities as discussed above and among others, includes quality control, inventory control, management, production planning and online changes in production plans. As these activities are diverse in nature and demand a range of perspectives to be solved efficiently, there is a need for coordination and integration of these various activities.

In general, operations management of batch process plant is considered more difficult than a continuous process plant. In continuous process plant operations, such as in a dry-process cement plant, there is one-to-one relationship between the processes and the process units and the relationship remains fixed over the period of time. In a continuous process plant, it is not necessary to separate process management and the unit's management. Whereas, in a batch process plant, operations and states of operations carried out in the plant change continuously, depending on scheduling. To operate a batch process plant efficiently, the plant information needs to be re-configured dynamically to correspond with the current operation status. This requires management of plant information re-configuration and separate handling of process management and unit management. This leads to complexity and need for modeling of operations. The available methods such as sequence function charts are used to express operations. SFC elaborates the control and manipulation of final control elements such as actuators, valves, pumps, etc. In SFCs, both the process information and plant structure information are embedded implicitly. If there is any change in plant structure or the operation procedure, it is better to configure a SFC from scratch than to modify an existing SFC. Further, the SFC configured is a single-dimensional sequence of manipulation, and lacks the ability to hold design rationale used at the time of design stage. It is not possible to use operation analysis and the real-time operation information.

The ISA S88 Part 1 standard is used to model the batch process operations clearly. As discussed in Chapter 9, the S88 standard defines four recipes and layered structure of operation. The management function is also modeled as a multiple-layer structure. The standard has a shortcoming, that separation of the plant information re-configuration management, process management and unit management is ambiguous.

10.3.1 Handling problems in batch production

Any batch control/management system must take account of what to do if the process goes wrong. The ISA S88 standard introduces S88 safe state, a structured approach to abnormality handling. Any batch process typically exists in order to manufacture a product. For example, in a batch reactor the sequence of events contain a series of individual tasks such as:

- Clean and check the empty vessel
- Charge with reactants
- Heat
- Cool
- Drain.

As per S88 standard, each task is known as a phase. A phase must contain the detailed instructions necessary for that task to be completed. For example, the heat phase might start by setting the temperature controller set point to 750 °C, checking the temperature until it reaches that value and then starting a 25 min timer. Before the S88 standard was available, such a vessel would be typically controlled using a sequence, programmed individually, and containing all the instructions necessary for the complete batch. Today, a much more structured programming technique is available in which the control structure is created using a building block approach which is usually based on libraries of components (such as charge or heat) and which is defined graphically by the engineer. This has obvious efficiency benefits, including reuse of code, reduced testing time, easier validation, etc. But one problem, which must be addressed by any batch control system, still remains. What to do if the process goes wrong? In the control systems before S88 standard evolved, it was not uncommon for majority of the code to be concerned with abnormality handling rather than the normal production tasks. If a problem were detected during a sequence, it would trigger the start of another sequence, which had the purpose of handling the problem. It was then the responsibility of the fault-rectification sequence to transfer control back to the primary sequence when the fault was rectified. This is obviously a complex programming task. And all of this structure, which on a typical batch plant involved numerous vessels and considerable complexity, needed to be designed, coded, tested and documented from scratch.

One of the best features of the S88 standard is its in-built structure for incorporating abnormality handling into the overall batch processing structure. Each phase contains the logic required for both the normal control of the sequence and the detection and rectification of an abnormality. Under normal operation, control processing cycles around the three states – idle, running and complete. Idle and complete are steady states, as they contain no instructions. Running is an active state and contains the sequential function chart. Apart from the above there are two other components. First, it shows the other states, which must exist to fully handle a problem. The batch may be placed in any of the steady states, stopped, held or aborted. The instructions necessary for achieving these states are held in the remaining active states: aborting, restarting, holding and stopping. Second, the failure monitor, who contains the logic, required for moving control to one of the error-handling active states. Each of these may or may not contain any processing instructions – that is part of the individual plant's batch control requirements – but the important thing is that the structure exists and can be added to in the future without modifying the primary processing logic. This defined structure, which allows for both normal production and abnormal situation control, is often termed safe state. It provides not only an industry-standard way of representing and structuring complex batch operations but also great benefits in efficiency of configuration, documentation, testing and validation.

10.4   Implementation of safety issues in batch management

Batch processes pose their own challenges in implementation of safety issues. Various processes have different kinds of risks, such as safety hazards, environmental risks, community risks and economical risks. One or all of these risks may require installation of a safety-instrumented system (SIS) for the process to reduce the risk(s). SIS are designed to respond to conditions of the plant, which may be hazardous and if no action were taken, could eventually give rise to a hazard. The SIS generates the correct outputs to prevent the hazard or mitigate the consequences.

The available batch standards do not address safety issues. However, safety standards for general process industry apply to the batch processes too. A SIS installed in a batch process may look identical to one installed in a continuous process. There are a number of other problems that must be considered while installing a SIS for a batch process plant.

As such, one makes no real distinction between safety in batch processes and safety in continuous processes. There are guidelines and procedures regarding safety that apply to both continuous and batch processes. Some of these guidelines and procedures have sections that are relevant only to batch processes, and these address batch specific needs or problems.

Following are some safety-related issues and practices followed:

- *Segregating the batch process control system and SIS*:   Basic process control systems do not achieve significant risk reduction. International standards such as IEC 61508/IEC 61511 indicate that the SISs must provide significant risk reduction. So, is a batch process requires additional risk reduction, besides the batch process control system there may be a need of installing a separate SIS. Separating the control from the safety in safety-critical processes creates an independent layer of protection. Such separation of the process control system and SIS into independent layers of protection is a fundamental and widely accepted practice in processes where safety issues are critical.

  In case of batch process control systems, the need for creating independent layers of protection by separation causes complexities, which are typically not found in case of continuous process for various reasons. For example, the SIS functions in case of batch process are recipe-dependent or even batch step-dependent. Hence, the SIS needs to know the recipe and the state of the batch process. This leads to two other major issues in batch processes. First issue is related to the setting of the recipe parameters in the SIS. And the second issue is related to the synchronization of batch states between the batch process control system and the SIS. In industry, practices followed for the separation of control and safety are no different for batch processes than for continuous processes. The solutions applied in the industry show a separate process control system and SIS, both the systems connected to the same field instruments. This practice is justifiable only when the SIS has the final verdict over the output instruments.

- *Synchronization of process steps*: As discussed in previous section, the need for separation of control and safety to provide an independent layer of protection leads to another issue in case of batch processes. The major issue in batch process control is the synchronization of the step sequence between the process control system and the SIS logic solver. The synchronization is a must, as the SIS logic solver needs to know the active recipe and the current batch step. As the process trip points are recipe- and step-dependent, adjustment of the trip points is closely related to batch process control system and SIS synchronization. Following are the two important practices concerning synchronization between the batch process control system and the SIS:

    - The recipe and batch step transmit from the batch process control system to the SIS. The SIS performs plausibility checks on the correctness of the step change conditions. An example of such a plausibility check is verification if the sequence of phase changes is plausible.
    - The SIS calculates the process step independent from the batch process control system. In other words, the SIS has to determine the batch step or recipe on its own measurement of process parameter, valve positions, levels, etc. As this sensing of the status of the batch process is safety related, it either enables or disables safety-instrumented functions. Hence, the batch status sensing should also be part of a safety integrity level (SIL) assessment.

- *Implementation of recipe-dependent variable trip points*: Implementation of recipe-dependent variable trip points in the SISs is another important safety issue in batch processes. Batch processes have an inherent characteristic – the sequence of various batch steps, each with its own properties and process safety limits. Because of these recipe-dependent safety limits, the trip levels in the SISs need to be adjusted with changes in recipe and the process step. It may also be necessary to change the logic for interlocks in the SIS, depending on the recipe selected.

10.4.1 Most critical trip point

There are a numerous solutions used in practice to cope with this problem. In a case where the equipment loss was the major concern but product loss was no concern, the solution is to establish a most critical trip point related to the equipment that is set in the SIS. The recipe-dependent trip levels are implemented in the batch process control system. However, this solution is applicable only when there is no risk reduction requirements related to the recipe trip levels.

10.4.2 Recipe strings

Another practical solution concerning the variation of recipe-dependent trip point level is the implementation of a string-based compare. A table of recipe strings and their corresponding trip point levels are stored in the SIS. Based on the recipe string, the SIS determines the appropriate trip levels from the predefined table. In the case of a recipe change, the batch process control system sends a string with the recipe name to the SIS. The SIS compares the string received from the batch process control system with the strings stored in the predefined table and adjust the trip levels accordingly. Receiving the correct string is a safety critical function; it must undergo analysis to ensure a low probability of dangerous failure.

This practice is further expanded. The safety system displays the selected recipe, and the operator acknowledges the selected recipe. After the operator acknowledgment, the trip levels are changed. Safety assessment of this solution with just string-based compare might show that it is not sufficiently fault-tolerant against different postulated faults of the batch process control system. One may prefer any other practice with additional fault tolerance.

Another practice involves use of a multi-position physical switch. The positions the switch can hold correspond to the various trip levels – for example, low, high and OK. The trip levels are stored in a predefined table in the SIS. A specific switch position then corresponds to a specific table cell. When the switch position changes, the trip levels in the SIS are set accordingly.

This practice is similar to the string-based compare industry practice, except that the recipe mode is communicated to the SIS through the physical switch instead of a recipe string. Similar to string-based compare, an expansion of this practice is also possible. After changing the physical switch's position, the SIS displays the selected recipe, and the operator has to acknowledge the selected recipe mode. Only after the operator acknowledgment, the trip levels are adjusted. An example of physical switch application and string-based compare is illustrated in Figure 10.3.

10.4.3 Measure process pulse

This practice involves the position of the physical switch compared with a recipe mode variable that transmits to the SIS via the batch process control system. The SIS adjusts the trip levels only when both the switch position and the recipe mode variable match. This compare constitutes redundant means by diverse technology and is therefore a very appropriate and safe method to change the trip point levels in the SIS. This practice is an example of diverse redundancy. Any of the techniques that increase safety, redundancy, diagnostics and lower failure rates can be applied to batch process safety problems.

10.4.4 Product ID pulsing

Another practice that addresses variable trip points is the product ID pulsing. The batch management system declares the product in manufacturing and sends the appropriate recipe information to the batch process control system. This recipe includes product ID number consisting of two or more digits. The same number also transmits to the SIS. In the SIS the trip levels are set according to the product ID number. The trip levels originate in a predefined table in the SIS. Then the confirmation takes place to ensure that the same product ID number is in both the SIS and the batch process control system. This practice is again similar to the string-based compare. After receiving the product ID number, the batch process control system starts pulsing two or more digital outputs to the SIS in a cycle that lasts until the next product recipe downloads. The SIS interprets these pulsed signals as the product ID number and compares them with the value sent directly from the batch management system. If the two values match, the various block valves controlled by the SIS can energize as the interlocks permit, else the safety system activates the interlocks, resulting in a safe shutdown. The trip levels selected by the SIS can be displayed for operator confirmation.

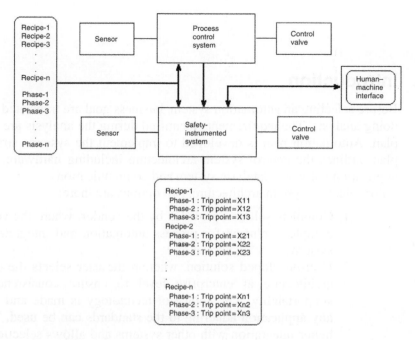

Figure 10.3
Example of trip point implementation using physical switch and string compare

It is important to note that whenever there is a discrepancy between the process control system and the SIS, the SIS should switch to the most strict and most safe, trip point levels. The safety system should always overrule the batch process control system.

11

Batch control technologies

11.1 Introduction

Before selecting an automation system, business goal are translated to systems strategy by doing analysis. The requirements identified during the analysis are part of the automation plan. Automation plan is developed to implement the systems strategy. The automation plan outlines the control system architecture including hardware, operating system and application software, database system and communications.

For selecting system architecture, two options are there:

1. Complete solution provided by the vendor, where the vendor chosen provides complete solution for process automation and integrates it with the business system.
2. User-developed solution, wherein the user selects the components of system architecture at enterprise level to ensure consistency, compatibility and supportability. The choice of technology is made and standards are set, and any application adhering to the standards can be used. This approach ensures better integration with other systems and allows selection of technologies that fulfill the business needs of the enterprise.

11.2 Overview of DCS/PLC architecture

11.2.1 Programmable logic controllers (PLCs)

The PLCs were initially developed and used for discrete operations as a replacement for electro-mechanical relays. However, over the period of time with addition of process control and networking capabilities, PLCs have gained greater functionality. With advancement in user interfaces and SCADA systems, PLCs have become more popular for sequence interlocking and integrating controllers.

PLCs are used with Personal Computers for batch management/control applications. Application of PLCs for batch control application is explained with help of an example.

Example 11.1:

Let us consider an example of automation of fermentation process. The batch process to be controlled consists of fermentation process for manufacturing anti-tuberculosis drug. Typical control system architecture for batch process control is shown in Figure 11.1. The Automation system consists of SLC 5/03^{TM} PLCs, panel view 550^{TM} operator interface, RS Batch^{TM} and RS View^{TM} software, and SCADA, for operator workstations.

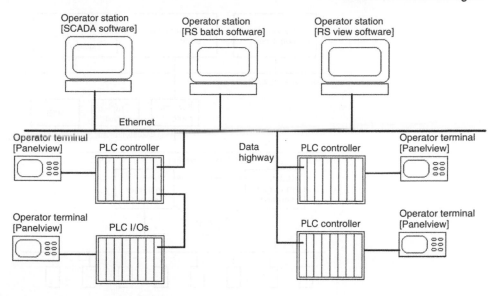

Figure 11.1
Example of PLC-based batch control system architecture

The PLCs control various process parameters such as the dissolved oxygen, temperature, pH, agitator speed, etc. to achieve the correct mix required for the batch. PLC I/O racks are mounted is remote I/O racks. All sequential control and interlocking is implemented in the PLC system. The PID controls are also implemented in the PLC system, such as control of the process water temperature, etc.

The panel view 550-operator interface for the operators and a supervisory control and data acquisition (SCADA) system for the supervisors provide them the view of the operation of the complete plant. The panel view 550-operator interface provides the operator facility to operate the system in automatic or manual mode, to set and monitor process parameters through the dedicated buttons configured for the specific purposes. Alarms can be viewed and acknowledged from both the panel view screen and SCADA system station. Process parameter logging, trending and report generation functions are built into the SCADA system.

The above example only illustrates a typical configuration of PLC system architecture for batch control application for a fermentation process. Various system components with trade marks (™) are used for the purpose of illustration only.

11.2.2 Distributed control systems (DCS)

Distributed control systems (DCS) were introduced in 1970s to replace control panels. DCS enabled centralization of tasks but spread the risk of failures by distribution of functions. The systems offered easy configuration instead of programming to implement a control function. Today, DCS is available with capabilities of discrete and sequential control and information management system. Application of DCS for batch control application is explained with help of an example.

Example 11.2:
Let us consider an example of batch control management system executed with a DCS architecture for a large scale biotech batch manufacturing plant. The system architecture is shown in Figure 11.2.

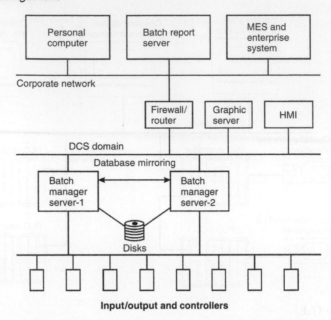

Figure 11.2
Example of DCS system architecture for batch control application in a biotech batch manufacturing plant

At the center of the DCS is a redundant pair of Compaq/DEC 4100 Alpha servers and a disk array configured in a true-cluster. The machines are cross-strapped with dual 100 mb ethernet connections to communicate with controllers, HMIs and with each other. At any point of time, only one server is active and acts as a master and does the job of scanning and updating the controllers. Databases of both the servers are maintained and updated real time by the mirroring program. The batch management program and historic recording programs run only on the active server. They are dormant on the standby server and get activated only when the active server fails and the standby server takes over the control function and becomes the active server. There are eighteen MMI or HMIs located at the manufacturing shop floor which access the graphics and real-time graphic parameters for display from six graphic distributed servers.

All the control application codes such as control modules, phases, unit control, etc. are located in unit controllers, which are located with I/O modules in I/O rooms. Control language conforms to IEC 61131-3 standard such as 4-mation. Control recipes are executed on the alpha servers, which contain the process model and the process supervisor. Batch and historic trend data is off loaded and histories are stored on separate report servers to reduce the load on the DCS control system. The DCS is considered a closed system and no users on the corporate intranet are allowed in the DCS domain. A router is setup as a firewall between the corporate intranet and DCS domain for certain approved communications. A batch report server is located on a corporate intranet and is utilized to pull, under secure conditions, certain batch reports of the batch historian and route these to the requesting user. A similar server is setup to make available historical trends. A MEC/DCS interface is used to send information of materials used in the process to a MES system for material genealogy tracing and inventory management.

The above example only illustrates a typical configuration of DCS system architecture for batch control application for a biotech batch manufacturing plant.

A block diagram of generalized system architecture of a batch control system, irrespective of whether the control system is PLCs- or DCS-based, is shown in Figure 11.3. The block diagram of batch control system architecture illustrates various components or sub-systems and their interface.

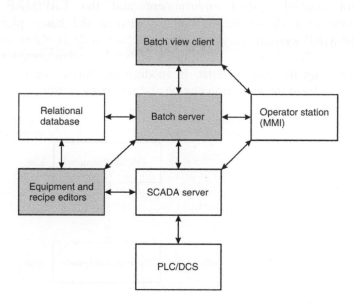

Figure 11.3
Block diagram of batch control system architecture

11.3 Integration of batch control systems to production management and ERP systems

Plant operations, in a modern batch plant, is a complex task as it includes a wide range of activities as discussed above, and among others include quality control, inventory control, management, production planning and online changes in production plans. As these activities are diverse in nature and demand a range of perspectives to be solved efficiently, there is a need for coordination and integration of these various activities, which can be addressed, by integration of batch control system with the production management and ERP systems.

We have a long way to go before all process plants have their control system integrated with the business management application layer. However, it is clear that the manufacturer who is able to respond quickly to changes in the business world by altering the manufacturing process will be well-placed to compete in the future. Most modern control systems are 'open systems'. They can be linked to other information systems much more easily so that technology is not the limiting factor as in the past. Today it is possible to send out a diagnostic alarm from a smart transmitter or smart devices connected through fieldbus through the control system onto the ERP system, which via e-commerce, can order a spare or replacement transmitter from the instrument manufacturer. Such integration was uncommon in the past. Today's technology allows very close integration between the process and business layers, but there is a huge amount of work required to define exactly what is shared between the two layers and how the data is used. The focus has shifted from the how to the why; instead of asking 'How can we make this system communicate', today enterprises ask 'What information shall we share and why?'

During the last decade, many companies have implemented ERP/MRP systems with the intent of achieving control over their global product supply mechanism. The main benefit of ERP/MRP systems is standardized data models that permit standardized interfaces to the plant floor. There are significant complexities and differences between the real-time plant control system environment and the ERP/MRP transaction environment. Figure 11.4 shows the difference between the batch plant control system and the ERP/MRP system. Any realistic interface reflects these complexities. The ISA S88 standard Part 1 provides definition of batch system complexity and the framework for interfacing the two systems. In modern enterprise control strategies, real-time control systems form the execution engine and reporting source.

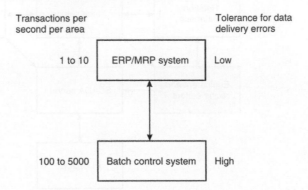

Figure 11.4
Difference between batch plant control system and the ERP/MRP system

Integrating batch control system with ERP/MRP systems poses a challenge, as the current ERP/MRP systems assume that the operator performs most operations and not the control system. Even some of the companies create internal standards to deal with data model; they find the integration of acquired manufacturing facilities a challenge. The difference between the plant control system and the ERP/MRP system can be brought down by use of ISA S88 Part 1 standard. The ERP/MRP system can comfortably operate at a general and master recipe level while batch control system can comfortably operate at the master and control recipe level. The main issue during integration is to distinguish between batch control events of enterprise-wide significance such as consumption of raw materials and events of local significance such as starting a pump and then to formulate an integration strategy around those events with enterprise-wide significance. Benefit of such ERP system is that it can be used as the intellectual property backbone for corporate-wide, global, recipe management.

Example 11.3:
Let us take a simple case of creating a control recipe based on information in a single ERP system to illustrate complexity in integrating an ERP/MRP system to a batch control system at manufacturing shop floor. Figure 11.5 shows typical steps required to schedule a single control recipe in batch control system based only on information in an ERP system.

The steps required to schedule a control recipe are:

- *Step-1*: The synchronization application waits for a recipe to be released to production within the ERP system.
- *Step-2*: The recipe information is downloaded to a transient location.
- *Step-3*: To specify a control recipe, the ERP system recipe header is used to query the batch control system for the required information.

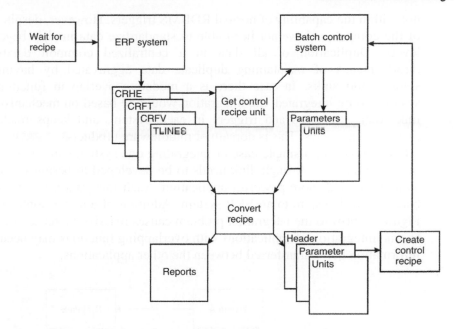

Figure 11.5
Example of creating a control recipe from an integrated ERP system

- *Step-4*: The batch control system returns the units to be specified to create a control recipe into a transient storage location.
- *Step-5*: The reports to be issued back to the ERP system at the completion of the control recipe execution are extracted from the ERP system recipe data and put into a transient storage location.
- *Step-6*: The batch control system header information, parameters and unit assignments are extracted from the ERP system recipe data.
- *Step-7*: The batch is scheduled in the batch control system using the data extracted from the ERP system.

To create a control recipe from the ERP system the following are required:

- Tables to store transient data
- Methods for data movement
- Encapsulated systems
- Dynamic data filters.

To keep the example simple, other important issues such as error checking, error recovery, redundancy, a third system has been ignored. The regular integration issues of data typing, data renaming, security contexts, etc. are implicit in this example, which illustrates challenges in integration of ERP/MRP system with a batch control system.

11.3.1 Integration problem domain

The main difference between integration and connection is that while integration the inherent functionality of each system must be considered. In case there is high overlap between the functionality of each system, the problem of integration can be reduced to a common database accessed by all the systems. The problem with the common database solution is that all applications must be designed from the ground up to use the common database format. The large amount of transient data and the complex triggers required are not within the capability of normal RDBMS triggers. However, despite many efforts most of the companies have not been able to standardize data models beyond the department levels. Duplication of all data in a centralized common database has its own disadvantages of maintaining duplicate data aggravated by having distributed data sources and sinks. In case there is a moderate overlap in functionality between the systems to be integrated, the integration would be based on mechanism that dynamically links data structures and triggers in each system, and keeps track of transient data management issues. This is domain of middleware products.

Let us consider a simple case of integrating one system with another system, as shown in Figure 11.6(a). A single link needs to be developed to permit the systems to function as a single and more powerful application. Such integration represents integration of a batch control system to an ERP system. Addition of another application(s) as shown in Figure 11.6(b) to the integration problem causes massive increase in the problem domain as true integration of applications with overlapping functions may need to know about the information being transferred between the other applications.

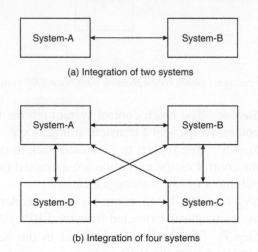

(a) Integration of two systems

(b) Integration of four systems

Figure 11.6
Integration of systems

Example 11.4:

An example of real integration problem in a plant with applications having overlapping functions is shown in Figure 11.7. A control recipe in the batch control system by an ERP system triggers the generation of sample IDs in a LIMS system. The problem of integrating a single batch control system with a single ERP/MRP system has many solutions but in practice it is not common because most of the real plants have peripheral systems that must be accessed as a part of batch control problem. In a realistic production operation, a general purpose, many-to-many integration solution is required to integrate a batch control system with ERP/MRP system.

11.3.2 Integration solution domain

For integration of a batch control system with an ERP system, examination of the integration problem domain leads to a simple set of criteria for the solution domain. The solution domains must consider the following:

- Integration of many-to-many systems
- Management of significant transient data storage

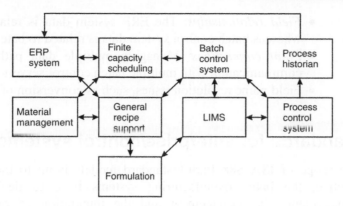

Figure 11.7
Real integration problem in a plant with applications having overlapping functions

- Many triggers based on data, events and other triggers
- Distributed operations across multiple computers and operating systems
- Error logging and recovery.

The integration solution for Example 11.4 is illustrated in Figure 11.8. The solution is chosen after considering the minimum number of interfaces to be supported by each system, apart from other reasons. As shown in Figure 11.8, each system could be integrated by developing a single interface to an Objectland. Once each application has been encapsulated as an object, all other integration issues could be handled in the middleware products. For example, VisualFlow™ or Microsoft Biztalk™, etc. could be used.

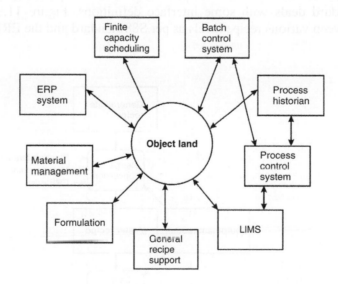

Figure 11.8
Integration solution with applications having overlapping functions

The middleware product handles following integration issues outside of all packages being integrated:

- *Native data typing including floating point differences, word size difference, processor differences, etc.*: The ERP system can run on computer systems but most of them are not same as the one used at the manufacturing shop floor.

- *Field relationships*: The ERP system data is related to the production order whereas batch control system data is related to batch ID.
- *Point connections*: Managing multiple data paths to ERP system and the communication with multiple batch control servers.
- Field name resolution issues, such as conversion of data names, etc.

11.4   Standards for enterprise/control systems integration

The scope of ISA S88 Part 1 standard models is up to the process cells. Whereas in practice, the batch manufacturing systems have to deal with the issues of area management, site management and the integration of batch manufacturing control systems with the other business management systems. The ISA S95 Part 1 standard on enterprise/control system integration contains models for the functions of area and site management and the interfaces between the enterprise business systems and the manufacturing control systems.

11.4.1   Benefits of ISA S88 standard to the ERP/MRP systems

The ISA S88 Part 1 standard outlines separation of the equipment model from the procedure model. The standard is useful for defining an equipment model that permits the definition of many recipes that can be operated in a single plant. The master recipe defines the steps to be performed by a specific grouping of equipment in an equipment-independent manner. The combination of the master recipe with specific equipment model creates a control recipe that controls the production operation. The S88 Part 2 standard deals with some interface definitions. Figure 11.9 illustrates the relationship between various recipe levels as per S88 standard and the ERP/MRP system.

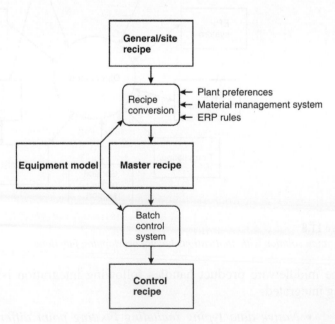

Figure 11.9
Relationship between recipe levels and ERP/MRP system

The benefits of ISA S88 Part 1 standard to the ERP/MRP systems can be summarized as:

- Standard batch automation methodology
- Standard road map
- Rapid recipe deployment
- Protection of intellectual property
- Reduction in startup risks.

The main advantage of a general recipe is the ability to specify a single recipe that can be run in many plants. It contains all the information required by the ERP/MRP functions such as material usage, major processing steps, etc. At the same time it does not contain minor details that are not significant for enterprise integration. The complex algorithms to convert general recipe to site recipe, and site recipe to master recipe are solved by a few companies only and are not recognized by the ERP/MRP systems.

11.4.2   ISA S95 Part 1 enterprise/control integration standard

The ISA S95 Part 1 standard describes models and terminologies for defining the interfaces between an enterprise's business systems and manufacturing control systems. The standard provides a common terminology, and a consistent set of concepts and models for integrating the control systems with the enterprise systems.

11.4.3   How does the ISA S95.01 standard help in systems integration?

Integration of manufacturing control systems with the business systems has been one of the difficult problems, not due to technology problems but due to people and organizational problems. There is a cultural difference between the information technology organizations and the people who develop and maintain the business systems, and the organizations and people who develop and maintain the manufacturing control systems. There are same terms often used for different things by these two groups and also there are different terms used for the same thing. This makes the problem of integration more difficult.

The ISA S95 Part 1 standard offers a solution by defining a common set of terms and definitions of the information and activities associated with logistic and manufacturing integration. It provides a common dictionary of terms that are useful during integration problem-solving. The terms include definitions of the activities of business logistics systems, the activities of manufacturing control and coordination systems in multiple levels of details and the information, which must be exchanged between these activities.

The standard comprehensively details the interface content between manufacturing control functions and other enterprise functions, with the intent of reducing risk, cost and errors associated with implementing these interfaces. The standard defines information exchange that is robust, safe, cost-effective and preserves the integrity of each system information and span of control. The ISA S95 Part 1 standard reduces the need for custom integration solutions, simplifies multi-vendor integration, and improves reusability and transportability of functions enterprise-wide. The standard may be used to reduce the effort associated with implementing new product offerings, with the objective of facilitating the development of business and control products that interoperate and easily integrate.

The S95 Part 1 standard starts with a definition of the domain of manufacturing control and general activities in the manufacturing domain. The standard contains a model of the functions with a manufacturing enterprise that relate or interact with the manufacturing control functions. The functions that are directly related to the scope of the standard are given additional definition and description, and the information flow between these functions is defined. The information flows are divided in to the related sets of information. Finally the detailed definition of the information and its internal relationships are modeled. The standard uses multiple models to describe the elements of enterprise/control systems integration. The initial models in the standard are very abstract but the final models are detailed and specific in nature. Each model adds a level of detail and definition, and builds on the information from the previous model.

11.4.4   Enterprise/control domains

The standard defines the control domain, and anything not in the control domain is defined as being in the enterprise domain. The activities/functions are included in control domain based on the following criteria:

- The activity/function is critical to plant reliability.
- The activity/function is critical for maintaining the regulatory compliance such as safety, health, environmental and current good manufacturing practices.
- The activity/function impacts the operation phase of the facility's life, as opposed to the design and construction phases of the facility's life.
- Also included is the information that the operators need to know to perform their tasks. Even though the activities generating the information are outside the control domain, they are defined in the standard as the information is required in the control domain.

11.4.5   Hierarchy model

In the S95 Part 1 standard, the initial hierarchy model is the most abstract model. It provides a hierarchical view of the activities associated with the manufacturing enterprises. The model is a simplified version of the Purdue Reference Model for computer-aided manufacturing, combined with the manufacturing execution systems association (MESA) model for the activities in the manufacturing control domain. The S95 standard does not define the levels 0, 1 and 2 of the hierarchy model, as these are defined by the other standards. ISA S88 Part 1 standard defines batch control and the S88 Part 2 standard deals with some level 2 to level 3 interface definitions. The S95 Part 1 standard deals with the level 3 to level 4 interface of the hierarchy model. The activities in level 3 are based on MESA model and are related to general areas such as data collection and acquisition, process management, production planning and tracking, performance analysis, operations and scheduling, quality management, resource allocation and control. The standard also defines activities where the enterprise and the control systems share the responsibilities. This includes areas such as document control, labor management and maintenance management.

11.4.6   Data flow model

The activities are given further definition and the relationships between the activities are modeled in a data flow model. The data flow model identifies specific functions in an enterprise and also defines how these functions interact with the process control function of manufacturing. The functions identified in the data flow model include:

- Production control management
- Production scheduling

- Materials and energy management
- Quality assurance
- Maintenance management
- Research and development, and engineering
- Procurement
- Product cost accounting
- Product inventory control
- Order processing
- Product shipping/dispatch management
- Marketing and sales.

The data flow model structure does not reflect an organizational structure of a company but it only represents an organizational structure of the functions. The data flow model depicts the boundary between the business systems and the manufacturing control systems. The standard helps in identifying where any particular function is performed in a company. The information detailed in the data flow model includes the following:

- Production schedule
- Actual production plan
- Production capability
- Incoming order confirmation
- Short-term and long-term energy, and material requirements
- Energy and material inventory
- Production cost objectives
- Production performance and costs
- Process data
- Process and product known how
- Maintenance requests
- Maintenance responses
- Maintenance standards and methods
- Maintenance technical feedback
- Quality assurance results
- Standards and customer requirements
- In-process waiver request
- Finished goods inventory
- Pack out schedule.

11.4.7 Categories of information

The information in the data flow model is 'objectified' or represented in form of a formal object model. Many of the data flows contain multiple objects and many objects exist in multiple data flows. The cross-reference between the data flows and the objects in the objects model is included in the standard. As the object models are very detailed and complex, it is difficult to use these to understand the general collection of information. For easy understanding, the information described in the object models is collected into information categories. The categories of information provide an overview for the object model. The information categories also provide a way to describe some new concepts and terms.

There are three main categories of information:

1. *Product definition information*:   How to make a product
2. *Production capability information*:   What is available for production
3. *Production information*:   What to make and results of making it.

Each of the above categories of information is further sub-divided into the information required by different functions. Some information sets overlap. For example, information related to production control, scheduling, maintenance, inventory, bill of materials and bill of resources. The overlapped sets of information are defined in the detailed object model. Areas that do not overlap or are not directly within the control domain are not defined in the model.

Object models

The ISA S95 Part 1 standard defines more than fifty different objects in the object model. The object model uses patterns for equipment, personnel and material. The pattern relates the object, such as a piece of equipment or material lot to its associated class. The classes contain the definitions of properties and their specific values. There is a quality assurance or qualification test that may be associated with each property value. The equipment, personnel and material definitions are used in segment definitions, in production request and production responses. Each object also has a definition of what it is and a list of information that it may include. For example, a material may include:

- An unique identification number for the material lot
- The quantity of material in terms of counts or weight
- The unit of measurement for the material in terms of numbers, kilograms, etc.
- The location of the material.

The object model may be used as the basis for formalized information exchange protocols, such as IDL objects definitions, SQL tables or XML files.

11.5   Sending process quality and production reports back to ERP

The use of ISA S88 Part 1 standard permits the automation of the reaction to quality issues as soon as the quality issue creeps up. Figure 11.10 shows an example of recipe branching on a quality issue.

Depending on quality test result, the recipe branch downs to a path that is not executed normally. Here, depending on the operator response to the quality check phase CHECK_PULP_SR, result will be either normal execution of the recipe (i.e. TRANSFER) or a branching of the recipe to an operator prompted phases (RECIRCULATE phase followed by REFINING phase). On completion of the branch path, the operator will be prompted for quality data again. Building such structure into the recipes at the startup permits the project team to share learning about reacting to quality issues in an effective manner. When the batch is not run manually to recover from quality issues there is a significant reduction in batch variability. This benefit can be noticed mostly during the project startups but also has a long-term impact on the overall effectiveness of the organization as learning and experiences from many sites are incorporated in the basic company recipes. This way the company's intellectual property is also encapsulated into the procedure, and saved and shared across multiple sites and operators. As a quality issue cannot be predicted in advance at the ERP level, it becomes the responsibility of the batch control system to return the material usage data not specifically called for at the ERP level.

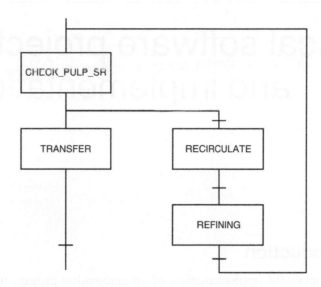

Figure 11.10
Recipe structure branching

However, with the eventual development of general recipe support at the ERP level, the reaction to quality issues will be deployable on a goal basis with the ERP level being the global repository of quality reaction mechanisms. Agile manufacturing will require such capabilities since shorter production runs will not have sufficient time for traditional site-to-site knowledge transfer mechanisms to take place.

12

Practical software project planning and implementation

12.1   Introduction

For successful implementation of an automation project, it is important to follow some simple concepts as following:

- Clearly define your automation objectives
- Solicit your experts
- Simplify and optimize
- Focus on value addition
- Adapt modular manufacturing principles
- Make your process communicate
- Retrain and realign
- Stick with it till it works
- Continuous evaluation and continuous improvement.

A successful automation project can help in meeting the business goals. So while translating business goals to system strategy, start with an analysis. The analysis should include what is the present situation and what challenges need to be met. The result of such analysis brings out the requirement. These requirements are part of an automation plan. The system strategy must also have an implementation plan.

A system strategy is formulated to address the following issues:

- What systems will be used?
- How the system is going to fit into the enterprise?
- What will be the approach for the life cycle of the system?
- How and when the new technology will be evaluated?
- Level of consistency required among the systems?

12.2   What to look for in batch software packages

Fundamentally, all batch plants process defined quantities of raw materials to produce the specific quantity, i.e. a batch, of finished product. How each plant does this depends on many things including the complexity of the process actions, the number of products the plant must produce, the differences between each product's process actions and sequences, the level of automation employed and the requirements for batch history. In many ways, all batch plants are the same but if we take a closer in-depth look at the batch plant we find many unique characteristics and requirements that give each plant its own personality. Knowing the personality of a batch plant is the key to selecting the right batch automation solution that best fits the plant's requirements.

Every batch plant has certain basic capabilities that are common to all batch plants. For selecting the most appropriate automation solution it is critical to identify and understand the dominant traits of the batch process. These include the process equipment, equipment control systems, operators who interact with the process, product recipes, a production schedule, a historical data record, etc.

12.2.1 Equipment control systems

Process/equipment control equipment comes in many forms and is usually dependent on the complexity of the process and the procedure that must be executed to produce that batch. The most basic is a fixed-speed mixer motor, with a starter and a timer that is set and controlled by an operator. Advanced process control solutions include PLCs, distributed control systems (DCSs), multi-loop controllers and some sort of human–machine interface (HMI) or operator interface.

12.2.2 Human–machine interface or operator interface

Operators must have a way to monitor and interact with the process. Human–machine interfaces or operator interfaces can be in form of a simple push-button panel or software displays on a CRT monitor. The human–machine interface serves many purposes, including process visualization, interfacing to batch and unit control, alarm annunciation and basic recipe management.

12.2.3 Product recipes

As discussed in Chapter 11, every batch plant has recipes for the products it manufactures. Every product recipe consists of a header, equipment requirements, formula and procedure. Every product produced in a batch plant has a recipe consisting of these components. In case of some products, they may not be obvious, but they do exist. For example, a recipe may be on a piece of paper that the operator uses to produce the batch. The recipe header and formula are defined on the paper. The equipment requirements and procedure is implied, e.g. blend all ingredients together in a blender. In a more automated process that has a control system, the formula values and procedure may be fixed in that control system. In some plants, the products to be produced all have the same procedure and only the formula changes. If there are only a few products, then formulas can be stored in the control system and the operator selects the product to be run by pushing a button or turning a selector switch to the desired product. Control system logic is used to switch-in the new formula values. When there are more products than can be stored in the control system, the capability to create, edit and manage recipe formulas is provided externally. Similarly, a facility that downloads new formulas to the control system may be provided when needed.

12.2.4 Production schedule

Every plant executes based on a production schedule. The schedule may come from a planning entity somewhere in the company and it is usually produced on a regular schedule such as monthly, weekly or daily. The schedule may be produced in different forms, such as a computer printout, a spreadsheet file or electronically downloaded to the batch execution system. The production schedule received from planning typically only addresses the production of products, not the other collateral production procedures that must be executed, such as the cleaning procedures that are commonly required in the food and pharmaceutical industries.

12.2.5   History

The batch plant must capture all data related to what has been produced in each batch. This may be done in both hard copy form and electronic archives. The most demanding history requirements are found in the pharmaceutical industry, where all batch events that occurred during production must be captured, including: all procedural events such as batch start, hold, restart and complete; recipe procedure execution events; process alarms; operator changes; operator comments; material consumption; material production; and trends related to key process variables.

12.2.6   Complex personality traits

Selection process for the automation of batch processes become more challenging when the batch process plant has complex personality traits. Automation of a single-product batch process can be quite straightforward. But when new requirements are introduced, the level of complexity increases dramatically. There are four significant requirements that have notable effects on the complexity of a batch automation solution. These are:

- The number of products to be made
- The procedural changes involved from product to product
- The need to produce multiple batches concurrently
- The complexity of each product's historical data.

12.2.7   Number of products

As the number of products to be produced increases, the greater the need is for recipe management and for flexibility in the control system logic. A process that will only be making a few products may not need a recipe management system, but a process that produces twenty or more must have some external facility to manage the multiple recipes. A good recipe management system is important when new product introductions are common or if product recipes are constantly being changed to enhance existing products. Modifying control system code to accommodate the changes can be time consuming and expensive, so it should be avoided wherever possible.

12.2.8   Procedural changes

When every product has a different procedure, process complexity increases dramatically. One solution that has been used is to change the control system logic for each product. This might be practical if product changeovers are infrequent, such as in long product runs. But if many product changeovers are the norm, such as for making products to customer orders, then a sophisticated capability for downloading and editing recipes and procedures must be used. Additionally, control system logic must be modular in nature and the solution must accommodate a sequencing engine that is easily re-configured for the product being produced.

12.2.9 Concurrency

Concurrency is the requirement to produce many different products at the same time, on the same production line. In the simplest plant, only one product is produced on a production line at any one time. In a more complex batch process, one process unit is being cleaned, the downstream unit may be finishing production of product X and the upstream may already be producing product Y. Keeping track of what product is in a process unit and when one unit can transfer to another unit can be a daunting task for operators. Concurrency requires sophisticated equipment management logic for equipment arbitration, allocation and release. Only one batch is allowed to own or allocate a process unit at any given time. Once a batch is transferred to a downstream unit, the upstream unit is released, allowing another batch to allocate the unit for its use. The automation solution in this situation must employ batch management and equipment arbitration rules to prevent equipment conflicts and potential product contamination.

12.2.10 Historical data

In any batch production process, it is important to have an accurate and complete batch history for many reasons. Proving to oneself as well as the regulatory agencies (e.g. US Food & Drug Administration (FDA)) that a drug was manufactured correctly is one driving factor for pharmaceutical companies. The drive for electronic batch records has produced the FDA 21 CFR Part 11 Regulation on electronic records and electronic signatures that allows batch history to be recorded electronically. This is helping to replace older hard copy records and signatures, which were previously the only FDA-acceptable batch records. Optimizing the plant is another important reason for capturing batch history. The first step to improving the performance of any process is understanding what the process is doing – and collecting a comprehensive history that captures all events facilitating this enhancement.

Hard copy is still acceptable for batch records with simple batch processes and in situations where a complete history is not a requirement. But when product procedures are complex and extensive, filling in the hard copy batch record can be a time-consuming, inefficient and often an error-prone approach.

12.2.11 Selecting a batch control/management system based on plant traits

In previous sections we have discussed about various traits of batch process plants that are critical in selecting a batch control/management system for a plant. Only after identifying and understanding the dominant traits of the batch process plant, one can select the most appropriate process control system. For example, for a batch process plant, the number of products to be made is a dominant trait. In this case a HMI solution that has a basic recipe management system for managing product formulas is the most appropriate. In case any other traits of the plant are also dominant for the batch plant, then an additional investment in custom software for the control system or the HMI may be required.

For a batch process plant, where concurrency is the dominant trait and there are no product procedure variations but there are many product recipes and historization of data is important, the most appropriate batch control/management system is one that provides facility for managing recipes consisting of unit recipes, equipment management logic for concurrency, and that captures data for batch history.

In case, for a batch process plant all the complexity or traits are present and are dominating, then the product procedural changes is considered as the most dominant trait. Presence of all the above-discussed four traits in a batch process plant increases the level of complexity. In such cases, custom solutions are very expensive and time-consuming to implement and maintain. In such cases, batch control/management systems that provide a configuration environment for creation of a control and information model that enables plant equipment to be used as defined in the product recipe are most appropriate. In such systems, the batch engine is critical, as it is responsible for execution of the product recipe, equipment management and batch history.

So for selecting an appropriate batch control/management system, it is important to understand the requirements of a batch automation, which involves understanding of the batch plant and its dominant personality traits.

In a practical and successful batch management system, the software must take care of variety of disciplines. Typical batch management system software architecture is shown in Figure 12.1.

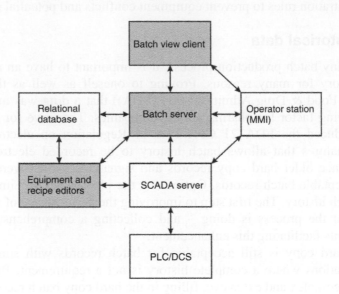

Figure 12.1
Typical batch management software system architecture

While evaluating or selecting a batch management software packages, following key capabilities and features must be looked for:

General

- It should be independent of process hardware
- It should be object-oriented
- It must be ISA S88 standard compliant
- It must have modular design approach
- It can be connected to any or existing process control systems – PLCs/DCS
- It should be software-independent
- It should be easy to integrate it with other business systems such as ERP/MRP systems

- Support available from the developer/supplier, for training, commissioning, maintaining and upgradation
- It should be based on the latest technology (it should not be obsolete) and the developer must support for the technology upgradation.

Capabilities for batch control/management

The batch control/management software package must provide following capabilities for effective control/management of batch process:

- Creation and management of master recipes
- Automatic execution of recipes
- Collection of batch process record data and generation of reports
- Process control through any PLCs/DCS
- Integration and exchange of batch and recipe information with corporate information systems
- Simulation of batch process
- Integration with wide variety of other complementary software applications.

The above capabilities/functionalities may be accomplished by the various components of the software package such as:

- *Equipment editor*: For configuration of physical equipment based on the physical model, such as areas, process cells, units, equipment modules, resources and unit tags, etc. Graphical tools for creation and maintenance of process equipment database.
- *Recipe editor*: For configuration and organization of recipe information – header, formula, equipment requirements and procedure for making a batch. The recipe information should be stored in a RDBMS format which can be modified outside the recipe editor when the system is integrated with the business systems – may be existing or in future. The software must provide graphical method such as sequential flow charts (SFCs) for creating and maintaining the recipes.
- *Operator interface*: The system must provide easy-to-use graphical user interface for operators and supervisors to control the batch process and functions of the system server. The operator interface should include security features for access and must provide platform for running other applications.

    The operator interface should provide following displays and features (note the list given below is only indicative and not exhaustive):

    - SFC graphical display of control recipe
    - Spreadsheet or tabular view of a control recipe
    - Batch list displays
    - Prompts for operator inputs
    - Phase control displays
    - Phase summary display
    - Resource allocation and arbitration displays
    - *Alarm displays*: Alarm summary, area-wise alarms, unit-wise alarms, etc.
    - Events journal

- Window for configuration of custom displays
- Help
- *Security features*:   Login based on password, and separate access levels for operators, supervisors and system administrators.

- *Server*:   The server is the execution engine; it integrates with process control systems (PLCs/DCS) and other software systems. The server performs various tasks of batch creation, execution of control recipes, equipment allocation, data collection and recording, etc. The server should also continuously take journal actions so that full recovery can be accomplished. To accomplish these functions, the server must support:

  - A bi-directional communication link with process control systems for issuing commands and receiving information in return
  - A communication link to seamlessly integrate with other components of the system and other software applications.

- *Archive*:   The system must archive the batch record. The archived information should be stored on any RDBMS that supports common interface standard such as Microsoft ODBC, etc. The user should be able to sort, analyze and generate custom reports using the information stored in the RDBMS.
- *Report editor*:   For generating reports that provide information about a particular batch, phase execution information, information on batch tracking, genealogy and other important parameters, and process performance indicators.
- *Simulator*:   A simulator is a power tool that provides facility of testing the recipes against plant configuration without actually running the plant. Simulators assist in debugging process during startups. The simulator should be easily configurable to meet the specific project or process requirements. It should be easy to modify or change phase states during running.

12.3   Batch control software products

To have a look on commercial batch control software packages available in the market, we will describe briefly key features of some software packages. The description of these products is based on the information available at the time of writing this book and may not include latest updates. The information on the products is used here only for academic illustration purpose.

12.3.1   InBatch™

The InBatch 8.0 batch management software package provides tools for automation for execution of production sequences and easy changeover from a product to another product. The InBatch batch management software has the following key features:

- Recipe management
- Plant modeling
- Tracking of bulk materials
- Scheduling of execution of batches
- Recording of batch, equipment and security history
- Schedule and view reports in a browser
- Option for a redundant batch server

- Facilitate system design as per 21 CFR Part 11
- Available in two editions
  - *InBatch premier edition*: For product recipes that involve different procedures and require flexibility
  - *InBatch flex formula edition*: For processes with variations between the products recipes is only in the formula. And execution of sequences never change.

The process sequence control and recording by InBatch are based on recipes. For example, the user can determine equipment-independent batch processes, without having to invest time and money in the development of separate control programs. InBatch controls the production sequence including dynamic batch and plant management as well as material management. Historic real-time data recording by means of an SQL database server and related flexible evaluation with various reports are sequence-supporting functions. Basic recipe creation is handled via a graphic editor in compliance with ISA standard S88.01. During the production runtime, the basic recipes are converted into control recipes, which start the process steps in the PLC system. Due to the open architecture, InBatch is able to communicate with other software systems. Via a defined interface, e.g. order data from an ERP system can be taken over into the batch control system directly by means of an interface module. The server–client configuration via ethernet TCP/IP permits decentralized access to the information from the batch management system. Via the various operating stations, the batch process is controlled according to requirements. Moreover, program editors for creation of recipes and the required engineering tasks are provided.

InBatch uses client–server system architecture. Typical system architecture of InBatch batch management system is illustrated in Figure 12.2.

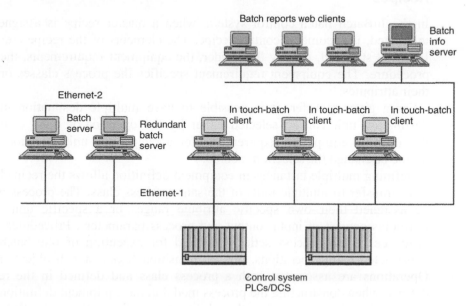

Figure 12.2
Typical system architecture of InBatch batch management system

Process, equipment and control

In the InBatch batch management system, the process is defined with units and connections between the units in accordance with the ISA S88 standard. The connection can be further divided into segments. Units with common functions or processing capabilities are grouped into a process class. All the connections between the same process classes are grouped into a transfer class. Process instances must be defined when a class is used in a recipe. Transfer instances are defined accordingly. The batch management system coordinates unit-to-unit material transfers during recipe execution.

The process classes exhibit processing capabilities, defined by the process phases and their associated parameters. Transfer classes have transferring capabilities, defined by the transfer phases and their associated parameters. Phase logic, which refers to the steps and sequences, executed in the control system can be constructed to accommodate automatically formula parameter values received during running.

The InBatch batch management system does the task of scheduling and execution of batches. Recipe is typically equipment-independent, as it refers to process and transfer classes only. The train is used to provide a list of potential equipments to the batch management system for dynamic selection during batch execution.

Process visualization with InTouch

For process sequence operation and monitoring, InTouch is used in the control room. This includes the display of all relevant process information, apart from display and operation of the InBatch batch management system, which are also organized by InTouch. For this purpose, an integrated ActiveX interface between InTouch and InBatch is provided, i.e. there is only one user interface, whereby a safety and identification system based on the LEGIC standard is used. Moreover, InTouch offers a central alarm signaling system and trend displays for analog measurement values.

Recipes

In the InBatch batch control system, when a master recipe is assigned to a train and initialized, it becomes a control recipe. The elements of the recipe are as defined in the ISA S88 standard, namely the header, the equipment requirements, the formula and the procedure. The equipment requirement specifies the process classes or unit classes and their attributes.

For a given transfer, it is possible to have multiple destination units. The desired destination unit can be selected either by the operator or done automatically. While defining the equipment requirements for a recipe, the unit selection mode is used to define or change the selection method.

Defining multiple instances in equipment definition allows the recipe builder to process in or transfer to multiple units of the same process class. The process instances can also be assigned their own specific attribute ranges or a specific unit can be assigned. Formulas consist of input, output and process parameters. Procedures which define the sequence of the process actions required for execution of one batch of a recipe are constructed using operations, phases, transition logic, branch objects and loop objects. Operations are associated with a process class and defined in the recipe. Phases are defined when constructing the process model in the equipment definition.

Transition logic provides the ability to re-direct the execution of a procedure depending on the result of a boolean expression. Branch objects are used to execute simultaneous operations and phases, execute one of the many operations or phases, and execute operations simultaneously on multiple units. Loop objects allow the recipe builder to re-execute the operations and phases, depending on an evaluated transition logic expression.

System architecture

The InBatch batch management system architecture consists of scheduling batches, initializing batches, coordination with the control system execution of batches, operator interface and storing all batch activities. This functionality is achieved through the batch manager, batch scheduler and batch display programs.

The batch manager directs execution of each batch. It coordinates usage of process units for each batch. The batch manager interprets recipes and enables the control system. Based on the recipe's procedure, blocks of control software and phase blocks are executed. Phase block control logic in the control system is responsible for controlling the process. The batch manager verifies each phase block whether it is ready for execution before it is enabled. If the phase block is ready for execution, phase values are downloaded to the block and the block is started.

The batch scheduler dispatches batches that are ready to run. Scheduling involves a manual entry of the batch identification, master recipe, batch size, and train into the batch scheduler. Batch initialization involves validation of the recipe, checking if the train exists, checking bulk materials, ensuring that the recipe's equipment requirements are met and verifying the compatibility of the process model database with the recipe.

The batch manager also interfaces with batch display programs. Through these displays, operators can put a batch or phase on hold and restart or abort batches or phases. Operators can also change phase parameter values, acknowledge the execution of phases, review phase interlocks status and enter comments while the batch is executing.

12.3.2 OpenBatch™

The OpenBatch 4.0 automation software provides support for NAMUR NE33 standard, which is precursor to ISA S88 standard for batch automation. OpenBatch provides a process hardware-independent, object-oriented, modular batch automation solution. The latest version of OpenBatch is designed to leverage Microsoft Technologies. OpenBatch 4.0 supports OLE for process control (OPC) communication protocol, which allows it to share information with other OPC complaint devices. The OpenBatch 4.0 software's ActiveX controls extend its capability to deliver a component based thin client and permits customized interfaces between human–machine interfaces (HMIs) and the batch servers. In other word, OpenBatch 4.00 can display its batch control and recipe logic inside other systems. Its cross invocation configuration allows other tools to be built and integrated back in the system resulting in reduced engineering time and simplified operating procedures.

OpenBatch provides following batch management capabilities:

- Creation and management of master recipes
- Automatic execution of recipes
- Collection of batch process record data and generation of reports
- Process control through any PLCs/DCS
- Integration and exchange of batch and recipe information with corporate information systems
- Simulation of batch process
- Integration with wide variety of other complementary software applications.

Process, equipment and control

In the OpenBatch, a batch process is modeled as areas, process cells, units, equipment modules and control modules in accordance with the ISA S88 standard. In addition to unit classes, process cell classes, equipment module classes, control module classes are introduced. The concept of equipment entity or equipment class as a collection of equipment that has essentially the same capabilities is well-defined in the OpenBatch.

In OpenBatch, for batch equipment definition, it is necessary to define at least one process cell less to a blank template. Then only the execution precondition information called arbitration information for each process cell instance is defined. A list of required resources must be available prior to starting any procedure on this specific process cell. This way, one process cell class may describe several instances, each dealing with raw materials supply, etc.

In the OpenBatch, the choice of specific unit instances is done in a similar manner as the configuration of process cell instances. In addition to unit name and arbitration information, associated data tags corresponding to the control system data points are also defined. While adding units, it is also possible to define inter-unit flow paths graphically.

The smallest equipment entities, equipment modules and control modules are defined by configuring equipment phases by the OpenBatch equipment editor. These are then provided as context-sensitive selections for recipe phases during the recipe configuration described subsequently. Phases are usually the logic within a control system, which performs the control actions. In OpenBatch, phases may also be called PC-bases phases, defined for execution by any PC on the network instead of the control system.

Recipes

The OpenBatch recipe editor complies fully with the ISA S88 batch control standard and supports IEC 1131-3 symbols and SFCs. Recipe procedures, operations and phases are identified as a part of procedural recipe. A one-to-many relationship exists between recipe phases and equipment phases. Phase control strategies in the OpenBatch simplify recipe creation and make equipment programming more intuitive by allowing the user to define more than one mode of operation for a single phase or a step in a recipe.

The OpenBatch recipe hierarchy – procedures, unit procedures and operations uses SFC to represent all levels of procedural functionality. At the apex level, only connections between unit procedures are described within the recipe. Each unit procedure may be opened to reveal its operations, which in turn may be opened to reveal their embedded phases. Looping and parallelism are likewise permitted at every procedural level.

System architecture

The OpenBatch batch management system software architecture consists of seven sub-systems or components as shown in Figure 12.3.

OpenBatch can be integrated with other automation systems as an OEM product to provide the required batch control functionality. Apart from equipment editor and recipe editor, another sub-system 'View' provides an operator interface for communication with the OpenBatch server. Many components of the 'View' are implemented as ActiveX control, and they can thus be inserted into container applications, notably web browsers.

From the functional point of view, it is interesting that using the 'View' the user can choose one of three allocation scenarios, namely:

1. Specify a unit to be allocated before execution
2. Prompt an operator for instantiating a unit from a unit class during recipe execution
3. Choose automatically the first available unit during recipe execution.

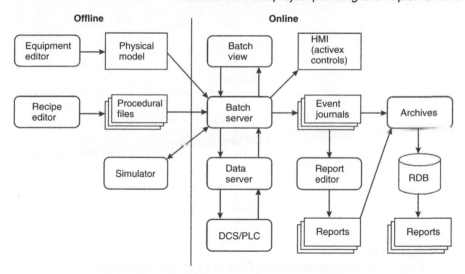

Figure 12.3
OpenBatch software architecture

The OpenBatch server is a centralized execution engine that executes the recipe and coordinates communications between the 'View', other sub-systems of the OpenBatch, process connected devices and other external software systems. The OpenBatch server also provides automatic restart control based on data logged in journals for all the actions taken so that a full recovery can be achieved in an event of control system failure. All the information available in the OpenBatch system is accessible through Microsoft component object model (COM), which makes it possible to integrate the OpenBatch with a wide range of other software system applications.

12.3.3　Plant Batch iT™

Plant Batch iT is a software package, which has been developed to satisfy the specific requirements of batch-controlled processes in the process industry. Its modular construction enables it to be adapted to widely differing scenarios and requirements.

Configuration

The Plant Batch iT software package provides configuration tool for parameterizing the batch process plant structure in a hierarchical manner, as illustrated in Figure 12.4. Specific parameters of the plant parts are determined here just like the individual basic functions of the plant. The result is a plant model, which exactly mirrors the technical circumstances and is the basis for creating and executing manufacturing instructions and control recipes.

Batch server

The Plant Batch iT server/recipe generator performs as system service central communication and coordination functions at plant level. An important function is the batch-related generation of control recipes and their transfer to the executing control system. By linking parts, list and process sequence the control recipes can be generated so that concrete plant situations, i.e. availability of operating materials and plant parts are taken into account. The batch server/recipe generator archives all batch data in the database.

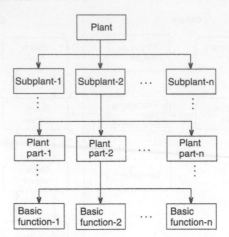

Figure 12.4
Configuration of plant structure in Plant Batch iT software package

Batch manager recipes

Plant-independent parts lists in terms of assignment of operating materials, quantity proportions, etc. can be created for products to be manufactured. The process descriptions/manufacturing instructions map the production engineering sequence. A graphical recipe editor supports the creation of process descriptions. Process sequences can be created and tested quickly and transparently by the user as a consequence of individual basic functions, and their linkages and dependencies.

Order list

The order list is an instrument for creating and progressing production orders and of representing in compressed form the current status of all planned, running and completed production orders.

Order and batch logs

All order and batch data are logged comprehensively in the database. Numerous selection possibilities simplify the selection of the data to be researched. A batch log contains all set points and actual values, starting and ending time, as well as the actual progress of the individual recipe steps.

Batch matrix/list

The batch matrix provides the user an overview of all batches currently running in the plant. The colored differentiation of different statuses of individual recipe steps enables fast orientation. It is possible to branch from here into the batch list.

The batch list visualizes all details such as status, set points and actual values, starting and ending times of the individual recipe steps of a batch. These details facilitate the necessary operating actions such as parameters and set point changes, aborting, stopping and continuing the batch.

Material management

This module undertakes the entire inventory control for all articles known to the system. Different storage locations can be managed. The batch system generates the stock movements automatically, e.g. raw material dosages and finished product receipts. Manual acquisition is also possible. Inventory corrections are possible through permanent or fixed-day inventory. The extensive statistics select and compress the movement data according to different criteria. The unbroken logging of the batch numbers for raw materials and also products enables batches to be followed up fully from the supplier of the raw material up to the final product.

Hardware requirement

Minimum hardware requirement for Plant Batch iT is a software package given in the following table.

System Requirements	
Server	PC Pentium III or higher, 256 MB RAM; Hard drive 9.1 GB Graphics card at least 1024 × 768 and 65535 colors
Workstation	PC Pentium III or higher; 128 MB RAM; Hard drive 4.0 GB Graphics card atleast 1024 × 768 and 65535 colors
PLC	Siemens S7-400, or any other compatible standard PLCs
Networking Requirement	
Network interface card	Intel Etherexpress Pro PCI-Bus network card or 3Com 3C509B ISA-Bus network card
TCP-IP	Ethernet network card
Software Requirements	
Operating system	Microsoft Windows NT 4, Service Pack 6a; German or English
Database	Microsoft Windows SQL Server V6.5, Service Pack 5a
	Language according to the operating system
	Language according to the operating system
Front end for logs	Microsoft Excel 97 or 2000 Language according to the operating system
Plant Batch iT requires the basic module Plant Direct iT	

12.3.4 RS Batch™

The RS Batch software package provides efficient, predictable operation of batch process plants and generates event information for the batch execution. It brings together all the facets of batch process automation and process management. It provides batch process management, which lets reuse code, recipes, phases and logic between phases within similar procedures.

RS Batch software package provides complete modular batch management solution. Keys features of the system are as follows:

- *Equipment editor*: The equipment editor defines the area model which represents processing capabilities of the physical equipment.
- *Recipe editor*: The RS Batch recipe editor uses SFC for recipe formulation. Each recipe is verified against the area model before it is released for production.
- *RS Batch View*[TM]: The RS Batch View provides an interface for creating, managing, executing and troubleshooting batches. It also provides critical information for operations.
- *Report editor*: The report editor also uses information from the RS Batch event journal and/or the database to create custom designed reports.
- *Archiver*: The batch information can be archived using the RS Batch event journal via open database connectivity (ODBC) to the database of user's choice.
- *RS Batch server*: The RS Batch server manages recipe creation and execution by controlling the process specific phase logic through phase transitions. The server also manages equipment arbitration, operator and external application requests, event journal recording and alarming.

    The RS Batch release 3.1 onwards, the RS Batch server and the archiver operates Windows NT[TM] services, which allows them to run in absence of an interactive Widows NT log on. Additionally, logging on or logging off the Windows NT system during operations does not disrupt the performance of the RS Batch server and the archiver. Both the RS Batch server and archiver can be configured to start automatically, giving control of the batch server to the service control engineer in Windows NT, and it can also be controlled manually using the RS Batch service manager. The RS Batch service manager can also be used to control the type of booting method that the batch server has to use and also configure the batch server to run in demo mode.

    RS Batch applications can support multiple language translations. This allows viewing and interacting in the language that is selected in the control panel. This includes language, date format, time format, list separator, number format, etc.

- *RS Batch HMI building blocks*: The RS Batch provides the ActiveX controls library that helps in creating a dynamic picture of batch control system data within the an human–machine interface. The ActiveX controls can be used to create custom-made interface. As the ActiveX controls library is designed especially for manufacturing and process control industry. Configuring the ActiveX controls is easy as it requires setting properties and does not require any code writing. The ActiveX controls library contains two controls that make up the RS Batch view components – control recipe list and the prompts list. These components equate to the batch list window and the acknowledged prompts window in the view component of RS Batch system.

12.3.5   VisualBatch[TM]

Process, equipment and control

In the VisualBatch, a batch process is modeled as areas, process cells and units. Further for each unit, the equipment phases are defined that execute on the unit. For each class of unit, several identical units determine a unit class. For each unit class, the class properties are defined and then a unit instance for each physical unit in the process cell is defined in an object-oriented manner by assigning values to the unit properties. The use of unit classes allows the users build class-based recipes.

Linking of units in the VisualBatch enables phases to communicate between the units. Linking units also defines the path or trains between units. Configuring of phases defines the equipment module on which the actual control logic is executed. VisualBatch allows user to configure equipment arbitration for handling multiple requests for the shared resource.

Recipes

The VisualBatch recipe editor complies fully with the ISA S88 batch control standard and supports IEC 1131-3 symbols and SFCs. In VisualBatch, unit procedures are also identified in addition to the identification of recipe procedures, operations and phases. When an operation is created in VisualBatch, a specific unit is assigned to it. When an operation needs to run on multiple units, class-based recipes may be used. A recipe can run on multiple units and can be assigned to any unit class at the time of running.

In VisualBatch, building recipes consists of completing the following tasks for the operations, unit procedures and procedures for master recipes:

- Defining the equipment requirements
- Creating sequential flow charts (SFCs)
- Creating recipe header
- Saving and verifying the recipe
- Releasing recipe to production.

In the VisualBatch recipe formulas can be created at any recipe level. Once a recipe is created, formula parameter values are set from the higher recipe level. Thus, when creating formulas for an operation, a unit procedure or a procedure and their values are set respectively from the unit procedure, the recipe procedure or by the operator when the batch is started.

System architecture

The VisualBatch is batch execution and recipe management software system. It uses Intellution's workspace for recipe development from a single location and integrates with SCADA and HMI packages. VisualBatch also integrates batch data and recipes into ERP systems. VisualBatch uses client–server system architecture. A typical VisualBatch architecture consists of VisualBatch server, one or more clients and a development workstation. In addition, for a SCADA or HMI system, the batch control system architecture can include one or more Intellution's FIX View clients and one or more SCADA servers.

The VisualBatch server coordinates the functionality of recipes, equipment database and each VisualBatch client during the execution. The VisualBatch server also generates batch event data and communicates with SCADA servers, the relational database and OPC or DDE process hardware. The VisualBatch supports one or more clients and servers. The VisualBatch clients also run FIX View, allowing the operators to monitor process values. The VisualBatch workstation is used for developing recipes and the equipment database. Using VisualBatch simulator, the automated batch process can be modeled and tested during the development stage.

Exercise 1

[A] Fill in the blanks, to complete the following statements:

1. Based on the output of the process, industrial processes are classified as _____ process, ____ process or _____ process.

2. Economy of scale, as a key to the success in business in chemical/process industries has led to the focus on designing and developing _____ processes.

3. Manufacturing of fine and specialty chemicals with the increased emphasis on high quality and customer requirements led to the focus on _____ processes.

4. Based on structure, batch processes are classified as _____, _____ and _____.

5. A multi-grade batch plant produces products that are similar but not _____.

[B] Identify the following processes as either batch process, continuous process or discrete process:

1.	Assembly of watches	[ ]
2.	Soap manufacturing	[ ]
3.	Beer brewing	[ ]
4.	Cement manufacturing	[ ]
5.	Manufacture of anti-tuberculosis drug	[ ]
6.	Production of cars	[ ]
7.	Manufacturing toothpaste	[ ]
8.	Electricity generation	[ ]
9.	Paper manufacturing	[ ]
10.	Assembly of television sets	[ ]

[C] List three batch processes (other than listed above):

1.
2.
3.

[D] State these statements as either TRUE or FALSE:

1. Batch manufacturing plants are comparatively more robust than a continuous plant.   [   ]
2. Batch process manufacturing facility is easy to scale up depending on market demand and requirements.   [   ]
3. In a continuous process, each equipment operates in a single steady state and performs a specific processing function.   [   ]
4. Batch process plants can be made flexible and are suitable for producing special products with few equipment.   [   ]
5. A multi-product, network-structure batch process is the most complex batch process.   [   ]
6. The ISA S88 Part 1 standard is a compliance standard.   [   ]

[E] Arrange the following elements of the physical model as described in the ISA S88 standard in hierarchical order (top to bottom):

Area, Enterprise, Control module, Equipment module, Process cell, Site and Unit.

Solution:

Exercise 2

Identify and define physical models

[A] The physical model described in the ISA S88 Part 1 standard has the following elements:

Area, Control module, Enterprise, Equipment module, Process cell, Site, Unit.

1. List elements of the physical model covered by scope of the ISA S88 Part 1 standards.

2. Which element of physical model is a domain for a batch control system?

3. Which physical model elements may contain element(s) of same levels?

4. Physical model element for carrying out major processing activity:

5. Physical model element that performs basic control:

6. Physical model element for carrying out minor processing activity:

[B] Tick those you will identify as a unit, from the following:

1. Reactor   [   ]
2. Pump   [   ]
3. Mixing tank   [   ]
4. Ingredient storage tank [   ]
5. A refrigerator   [   ]
6. A washing machine   [   ]
7. A kitchen blender   [   ]

[C] State the following statements as either TRUE or FALSE:

    1. Physical model is collapsible.   [   ]
    2. Batching can occur without units.   [   ]
    3. Control modules are defined in the batch management system.   [   ]
    4. A process cell is usually independent of product.   [   ]
    5. A process cell must contain at least one unit.   [   ]
    6. A control module does not have to be part of an equipment module to be a part of unit.   [   ]
    7. Unit can contain more than one batch at a time.   [   ]
    8. Equipment modules execute portions of a recipe, but control modules do not.   [   ]

[D] Identify equipment modules (EM) and control modules (CM) from the following:

    1. Weigh scale   [   ]
    2. An inkjet printer for labeling   [   ]
    3. Hoist   [   ]
    4. A damper actuator   [   ]
    5. A sampler   [   ]

[E] Figure E2.1 illustrates an example of a physical model of an ABC Enterprise.

Answer the following questions based on the physical model of the ABC Enterprise:

    1. Number of process cells in the ABC Enterprise?

    2. Total number of units in the ABC Enterprise?

    3. Process cell(s) for which no unit(s), equipment module(s) and control module(s) are illustrated in the physical model?

    4. Unit(s) for which no equipment module is illustrated?

    5. Unit(s) for which no control modules are illustrated?

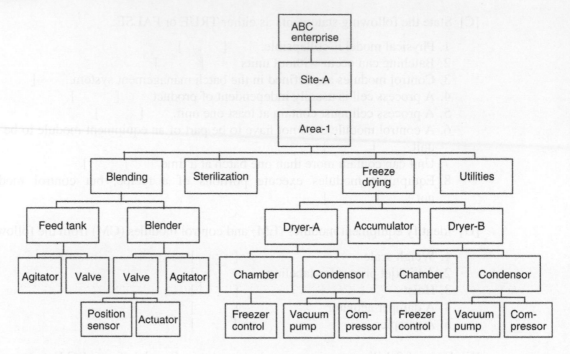

Figure E2.1

Exercise 3

Answer the following questions for the illustrated process model in Figure E3.1:

1. How many processes are in the model?

2. Identify the process operations in each process stage.

3. Identify the process actions contained within the process stages or operations.

4. Consider the model for fiber slurry and identify the process actions, process operations and process stages in the model.

Identify and define process models, actions, operations and stages

[A] List down the elements of process model in hierarchical order:

[B] State whether the following statements are TRUE or FALSE:

1. A process stage usually operates independently from other process stages. [   ]
2. A process operation describes a minor processing activity.   [   ]
3. In a process model, the procedure for making a product does consider the actual equipment for performing the various process steps.   [   ]
4. A process operation results in chemical and/or physical change in the properties of the material being processed.   [   ]

[C] Figure E3.1 illustrates an example of process model for raw material slurry batch preparation process for manufacturing fiber–cement sheets.

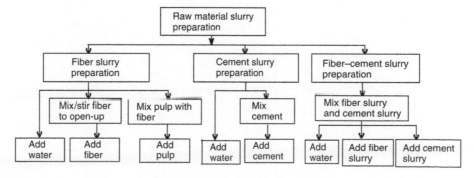

Figure E3.1

Answer the following questions for the illustrated process model in Figure E3.1:

1. How many process stages are there in the process model?

2. Identify the process operations in each process stage?

3. Identify the process action(s) common in all the process stages in Figure E3.1?

[D] Draw a process model for a batch process and identify the process stages, process operations and process actions in the model.

Exercise 4

Identify and define procedural models

[A] Fill in the blanks to complete the following statements:

1. Procedural control is a link between _____model and _____ model.
2. Two types of procedural elements are_____elements and _____elements.
3. _____defines the strategy for accomplishing a major processing action, like making a batch.
4. _____is a sequence of phases that defines a major processing sequence that changes the state of the material being processed.
5. _____control includes exception handling, interlocking, sequential control, regulatory control and alarm annunciation.

[B] State whether the following statements are TRUE or FALSE:

1. Procedural control involves control recipe procedure and equipment control. [   ]
2. Procedural control directs equipment-oriented actions in an ordered sequence to carry out process-oriented tasks.   [   ]
3. A recipe procedural element is independent of the equipment on which it is executed.   [   ]
4. More than one unit procedure can be active in a unit at a given point of time. [   ]
5. An operation usually involves a chemical or physical change.   [   ]

[C] Identify the blank elements (boxes) in the following diagram to complete the sequence of activities:

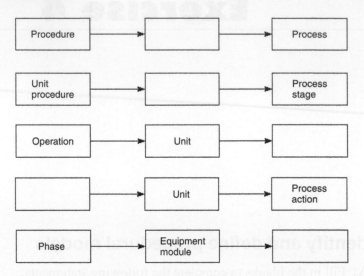

[D] Take an example batch process and draw the procedural model for the same consisting of at least one procedure, a unit procedure, an operation and a phase.

Exercise 5

Introduction to recipes

[A] List down the components of a recipe in the following diagram:

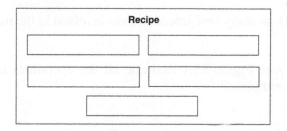

[B] Match the following:

1. General recipe   A. Accounts for specific information like equipment arrangements necessary to make the product

2. Site recipe   B. Contains complete scheduling, operational and equipment information

3. Master recipe   C. Describes the requirements to manufacture a product at various sites

4. Control recipe   D. Language, units of measurements and raw materials are adjusted

[C] State the following statements as either TRUE or FALSE:

1. A recipe should be independent from the equipment on which it is executed. [   ]

2. General recipe is created with specific knowledge of the process cell equipment that will be used to manufacture the product.   [   ]

3. Master recipe is targeted at a specific process cell.   [   ]

4. A master recipe can be created as a standalone entity without a general recipe or a site recipe.   [   ]

5. Usually, a recipe procedural element maps to an equipment procedural element at phase level.   [   ]

[D] Example of a typical recipe to manufacture toothpaste is shown in Figure E5.1. Answer the following questions based on the recipe.

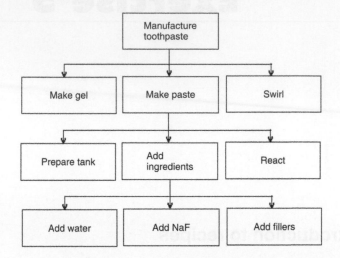

Figure E5.1

1. How many unit procedures are involved in the manufacturing process?

2. Does Figure E5.1 illustrate all the operations involved in the manufacturing of toothpaste?

3. Identify the phases shown in Figure E5.1.

[E] Take an example of a recipe manufactured using a batch process. Identify procedure, unit procedures, operations and phases for the same.

Glossary

A

ActiveX A defined set of technologies developed by Microsoft based on object linking and embedding (OLE) and component object model (COM).

ActiveX control ActiveX control is a standalone software component that performs a particular task in a predictable way. It can be used in many different applications, and the user will always know how to use it and how it is going to behave. ActiveX controls were formerly called OLE controls.

Alarm Audio-visual signal that indicates an abnormal or out-of-limits condition.

Algorithm A set of well-defined rules that specifies a sequence of operations for performing a specific task.

Allocation A form of coordination control that assigns a resource to a batch or unit. An allocation can be for the entire resource or for part of a resource (S88).

Application A software program written to perform a task on a computer.

Arbitration A form of coordination control that determines how a resource should be allocated when there are more requests for the resource than can be accommodated at one time. Arbitration is typically required when a shared resource is requested by more than one batch or operators (S88).

Archiving The process of moving historized data into a separate database than Historian database so that the portion can be managed independently, i.e. backing up onto an external medium, putting into offline state or back into online state.

Area A component of a batch manufacturing site that is identified by physical, geographical or logical segmentation of equipment within the site. An area may contain process cells, units, equipment modules and control modules (S88).

B

Basic control A type of control that is dedicated to establishing and maintaining a specific state of equipment or process condition. It may include – discrete or sequential control, regulatory control, interlocking, monitoring and exception handling (S88).

Batch Batch means both the material made by and during the process and also an entity that represents the production of that material. Batch is used as an abstract contraction of words 'the production of a batch'. A batch is a running control recipe (S88).

Batch control Control activities and functions that provide a means to process finite quantities of input materials by subjecting them to an ordered set of processing activities over a finite period of time using one or more pieces of equipment (S88). Also, batch control consists of a sequence of one or more steps or phases that must be performed in a defined order for finite period of time to process finite quantities of input material to produce finished product.

Batch ID   An unique identification name/number given by the operator to each batch.

Batch process   A process that leads to the production of finite quantities of material by subjecting quantities of input materials to an ordered set of processing activities over a finite period of time using one or more pieces of equipment (S88). Also, a sequence of one or more phases that must be performed in a defined order and results in finite quantities of material.

Batch report   A report generated on completion of a batch that details events that took place when the batch was running.

Batch schedule   A list of batches to be produced in a specific process cell. Also, the batch schedule typically contains information such as – what is to be produced, how much quantity to be produced, when or in what order the batches are to be produced and what equipment are to be used (S88).

Baud   An unit of signal speed, number of discrete signals per second.

C

Code   A computer program or a symbolic form of data representation that can be accepted by a computer.

Collapsing   Elements, levels or layers in the models may be omitted as long as the model remains consistent and the functions of the element removed are taken into account (S88).

Common resource   A resource that can provide services to more than one unit or phase. Common resources are identified as either exclusive-use resources or shared-use resources (S88).

Component object model (COM)   is a Microsoft-defined underlying architecture that forms the foundation for higher-level software services such as provided by OLE. A standardized way of linking software components, from possibly different vendors.

Computer system   A group of hardware components assembled to perform in conjunction with software programs designed to perform a specific functions.

Connections   Connections are defined as the equipment necessary for transferring a product from one unit to another.

Continuous process   A process in which material flows in and the product leaves the system uninterrupted for extended periods of time.

Control action   Control action of a controller or a control system is the nature of the change in the output effected by the input.

Controller   A device that operates automatically to regulate a controlled variable.

Control loop   A grouping of instruments, control algorithms and actuators, designed to measure and control a controlled variable.

Control module   is the lowest level grouping of equipment in the physical model that can carry out basic control (S88). Also, a control module consists of sensors devices and other control modules that together perform a specific task.

Control recipe   A type of recipe which through its execution defines the manufacture of a single batch of a specified product (S88). Also, a control recipe defines manufacturing environment for a single batch and include the specific equipment and raw materials to be used. Control recipe is derived from master recipes.

Control system   A PC or computer-based system that controls manufacturing plant operations. Control systems are real time, require high reliability and availability.

Coordination control   A type of control that directs, initiates and/or modifies the execution of procedural control and the utilization of equipment entities (S88).

D

DCS   Distributed control system is a process control system that employs multiple computer units or controllers at different locations in the plant.

Device   An apparatus for performing a prescribed function. A single physical piece of plant equipment that has an active function in the process. E.g. pumps, valves, etc.

Deferred parameter   Step formula value whose value is deferred to take on the value of another parameter at a higher level within the recipe. This function has the effect of passing recipe formula parameter data from one level of recipe down to another.

Distributed component object model (DCOM)   A Microsoft protocol that enables software components to communicate directly over a network in an efficient, reliable and secure manner. Software components that reside on physically distributed computers.

Dynamic   Continuously changing or updated.

Dynamic data exchange (DDE) server   Server that retrieves data from any NetDDE-aware process hardware. DDE is a form of communication that uses shared memory to exchange data between applications.

E

Element   A component of a device or a system.

Embed   Insert an object created in another application that supports OLE (such as Microsoft Word documents).

Embedding   A form of copying in which the copied object resides in the destination file only, with no link to the source file, but can be edited using the same tools available in the source file. Changes made to an embedded object exist only within the destination file and do not change the source file from which the object was copied. Likewise, changes made to the source file are not reflected in the embedded object.

Enterprise   An organization that coordinates the operations of one or more sites (S88).

Equipment control   The equipment-specific functionality that provides the actual control capability for an equipment entity, including procedural, basic and coordination control, and that is not a part of the recipe (S88).

Equipment database   The physical component of a batch facility. A database which consists of all equipment in the facility and all of the tasks that it is capable of performing.

Equipment entity   A collection of physical processing and control equipment and equipment control grouped together to perform a certain control function or set of control function (S88).

Equipment module   A functional group of equipment that carry out a finite number of specific minor processing activities (S88). Also, an equipment module consists of equipment and control modules that together perform a minor processing task or a phase.

Equipment operation   An operation that is part of equipment control (S88).

Equipment phase   A phase that is part of equipment control (S88). The logic for an equipment phase resides in the process control system.

Equipment procedure   A procedure that is part of equipment control (S88).

Equipment unit procedure   An unit procedure that is part of equipment control (S88).

Event   An asynchronous generation of a message caused by the change of state of some resource or process.

Event message   A message generated typically by the process control system, to inform users of an abnormal condition or of a significant process occurrence.

Exception   An exception is an event that occurs outside the normal or desired behavior. An exception can occur at any level in the control activity model.

Exception handling Functions that deal with the plant or process contingencies and other events which occur outside the normal or desired behavior of batch control (S88).

Exclusive-use resource A common resource that can be used by only one user at any given time (S88).

Expanding Elements, level or layers may be added to the models without affecting the integrity of the original relationship between the elements (S88).

F

Formula A category of recipe information that includes process inputs, process parameter and process outputs (S88).

G

General recipe A type of recipe that expresses equipment- and site-independent processing requirements (S88). Also, a form of product recipe which is equipment-independent and contains the processing actions required to produce a product without regard to any specific unit layout or process cell structure.

H

Header Information about the purpose, source and version of the recipe such as recipe and product identification, creator and issue date (S88).

HMI Human–machine interface (HMI) or man–machine interface (MMI).

I

ID An unique identifier for batches, lots, operators and raw materials (S88).

Initial step The logical start of a sequential function chart (SFC).

Interface A shared boundary. A hardware circuit or software that enables communication.

Interlock A device or control action designed to avoid hazardous condition by taking a predefined action.

L

Lot An unique amount of material having a set of common traits. Examples of common traits are material source, the master recipe used to produce the material and distinct physical properties (S88).

M

Manual mode A state associated with a step in a batch. When a step is in manual mode, its transition does not execute until an operator sends a message instructing it to do so. Also, the mode of a control recipe when the procedure does not sequence automatically. Transitions are not checked, therefore the step will perform its assigned functions and no further action will be taken without operator intervention.

Master recipe A type of recipe that accounts for equipment capabilities and may include process cell-specific information (S88). Also, a recipe that defines the equipment requirements to manufacture a product. This equipment is grouped into process cells. Master recipes are designed by control engineers to run on many different lines within a process cell.

Material transfer capabilities/phases Each transfer class, and therefore each connection in transfer class has transfer capabilities that are defined for transfer class.

Mimic A screen display that indicates the layout of a plant or process.

Mode The manner in which the transition of sequential functions are carried out within a procedural element or the accessibility for manipulating the states of equipment entities manually or by other types of control (S88).

Modular The term referring to a sub-division for standard assembly.

N

Network A series of devices connected by some type of communication medium. A network may be made up of several links.

O

Object A self-contained module consisting of related data and procedures.

OLE Object linking and embedding.

OLE database Microsoft-defined interface protocol and standard, which allows access to various types of source data in the form of structural two-dimensional array.

OPC OLE for process control; a defined set of interfaces, based on OLE/COM and DCOM technology.

OPC server OPC servers implement OPC COM objects and their interfaces. An OPC client can configure the rate at which an OPC server should provide the data changes.

Operation A procedural element defining an independent processing activity consisting of the algorithm necessary for initiation, organization and control of phase (S88). Also, an independent production activity within a unit procedure, consisting of phase names and the algorithm necessary for the initiation, organization and control of those phase names. There may be one or more phases within an operation that may execute sequentially or concurrently.

Operational specification A document that defines how an automated process will operate.

P

Path The order of equipment within a process cell that is used or is expected to be used in the production of a specific batch (S88). Also called as stream.

Phase The lowest level of procedural element in the procedural control model (S88). Also, a series of steps that cause one or more equipment- or process-oriented actions. These actions issue commands to set or change controller constants, modes or algorithm.

Phase state A term that declares the phase's current condition, such as idle, starting, running, complete, holding, held, restarting, aborting, stopping and stopped.

Phase logic interface (PLI) The interface between the batch server and the phase logic.

PID control Control action in which the controller output is proportional to a linear combination of the input, the time integral of input and the time rate-of-change of the input.

Procedural control   Control that directs equipment-oriented actions to take place in an ordered sequence in order to carry out some process-oriented task (S88).

Procedural element   A building block for procedural control that is defined by the procedural control model (S88).

Procedure   The strategy for carrying out a process. It refers to strategy for making a batch within a process cell.

Process   A sequence of biological, chemical or physical activities for the conversion, transportation or storage of material or energy.

Process action   Minor processing activities that are combined to make up a process operation. Process actions are the lowest level of processing activity within the process model.

Process cell   A logical grouping of equipment that includes the equipment required for production of one or more batches. Process cell defines the span of logical control of one set of process equipment within an area.

Process connected device (PCD)   The hardware that allows batch management software to communicate with the equipment in a facility.

Process control   The control activity that includes the control functions needed to provide discrete, regulatory and sequential control and to collect and display data.

Process input   The identification and quantity of a raw material or other resource required to make a product.

Process management   The control activity that includes the control functions needed to manage batch production within a process cell.

Process operations   A major processing activity that usually results in a chemical or physical change in the material being processed, and that is defined without consideration of the actual target equipment configuration.

Process output   An identification and quantity of material or energy expected to result from one execution of a control recipe.

Process parameter   Information that is needed to manufacture a material but does not fall into classification of process input or process output.

Process stage   A part of a process that usually operates independently from other process stages and that usually results in a planned sequence of chemical or physical changes in the material being processed.

Process validation   Process validation is establishing documented evidence that provides a high degree of assurance that a specific process will consistently produce a product, meeting its predetermined specification and quality attributes.

Process variable   Any variable property of a process.

Programmable logic controller (PLC)   A control device. Logic programs contained in the PLC-read inputs, run the program and write the outputs.

R

Recipe   The necessary set of information that uniquely defines the production requirements for a specific product.

Recipe formula parameters   Variables used to control process values such as time, quantities, temperature, etc. Recipe formula parameters let you create flexible and reusable recipes. Also, a parameter specific to a recipe that can be used to pass values from one level of a recipe to the next lower level.

Recipe header   Administrative information about the recipe. The information includes the procedure identifier, version number, version date and the author.

Recipe hierarchy   The procedural model described in the ISA S88 standard. The model defines procedures, unit procedures and operations in a hierarchy of recipes.

Recipe management   The control activity that includes the control functions needed to create, store and maintain general, site and master recipes (S88). Also, the process of creating, maintaining recipes.

Recipe operation   An operation that is part of a recipe procedure in a master or control recipe (S88).

Recipe phase   A phase that is part of a recipe procedure in a master or control recipe (S88).

Recipe procedure   The part of a recipe that defines the strategy for producing a batch.

Recipe unit procedure   A unit procedure that is part of a recipe procedure in a master or control recipe (S88).

S

Safety integrity level (SIL)   Safety integrity level is one of the four levels, each corresponding to a range of target likelihood of failures of a safety function. A SIL is a property of a safety function, rather than of a system or its component.

Sequence   The manner in which instructions are organized to be implemented in a device.

Sequential control function (SFC)   A graphical representation of a recipe.

Server   Computer-holding master files and data for distribution to other computers.

Shared-use resource   A common resource that can be used by more that one user at a time (S88).

Site   A component of batch manufacturing enterprise that is identified by physical, geographical or logical segmentation within an enterprise. A site may contain areas, process cells, units, equipment modules and control modules (S88).

Site recipe   A type of recipe that is site-specific. Site recipes may be derived from the general recipes recognizing local constraints, such as local language, available raw materials, etc. (S88).

State   The condition of an equipment entity or of a procedural element at a given time. The number of possible states and their names may vary for equipments and for procedural elements (S88).

State transition diagram/matrix   A diagram or matrix that illustrates the transition of procedural elements.

State transition logic   The logic within the PLI that provides a standard interface to the project-specific phase logic. The state transition logic receives commands from the batch server or the operator and then initiates the different components of the project-specific phase logic. It resides in the controller.

Step   Defines the logic of the recipe.

Structured program   A program constructed of a basic set of control structures, each one having one entry point and one exit point.

Supervisory control   Control action in which the control loops operate independently, subject to intermittent corrective action.

T

Tag   An entity that represents one piece of information in the process-connected device.

Train   A collection of one or more units and associated lower-level equipment groupings that has the ability to be used to make a batch of material (S88). Also, a generic name for a process line.

Transition   Defines when a recipe moves from one step to another step in the sequential function chart.

Trip point   A predetermined value of a system parameter that, when reached, causes an action to take place or a state to change. That action may be an alarm, a valve closure, etc.

Tune   To manipulate the configuration parameters of a loop in order to optimize its effectiveness.

U

Unit   A collection of associated equipment modules and/or control modules and other process equipments in which one or more major processing activities can be carried out. Also, a piece of equipment in a process cell that performs a specific task, consisting of all equipment and control modules.

Unit procedure   A strategy for carrying out a contiguous process within a unit. Unit procedure consists of contiguous operations and algorithms necessary for initiation, organization and controlling these operations (S88).

Unit recipe   The part of a control recipe that uniquely defines the contiguous production requirements for a unit. The unit recipe contains the unit procedure and its related formula, header, equipment requirements and other information (S88). Also the operations that control the function of a single piece of equipment.

Unit supervision   The control activity that includes control functions needed to supervise the unit and the unit resources (S88).

User interface   Programs that allow user interactions such as showing data, updating data and triggering functions to be performed.

Appendix A

Modular approach

A.1   Introduction

In past the process facilities were designed to produce a defined throughput of a specific product. Material balance equations and equipment selections were made to meet throughput requirements. Plant managers could be heard bragging when a facility was producing more than the design capacity. Seldom were calculations and designs developed for lesser throughputs. When market demands diminished and production rates were reduced, the oversized equipment, different process dynamics and control system tuning parameters made it difficult to produce a quality product. Similarly, when competition, regulation or market demands required changing the product, facilities often required extended shut downs and sizable investments to prepare the facility for the new product. Now a days, requirements for flexibility and time-to-market have become key business drivers. To quickly produce a new product, in customer-required quantities, producers are turning to batch processing and process automation. Agile is the buzzword-modular is the way to achieve it.

Companies face tough economic realities in the business of discovering and bringing new products to market. Only one in ten thousands of the new compounds discovered in a pharmaceutical lab survive for successful commercial production. When a new product is approved, it is imperative to maximize the return on investments by scaling up to commercial production quickly. This challenge can be only met using modular construction and design techniques. Although this can save time at the construction site, front-end design time may actually increase. Still the companies face the dilemma of being late to market or the risk of investing in new facilities before the new product is approved. This modular approach provides a method of reducing the risk by designing the building blocks of modern production facilities prospectively, well before new product approval is certain. Because many of the products currently in the development stage can be manufactured using similar core technologies, robust standardized process modules can be designed in advance, and taken 'off the shelf' when needed. The ISA S88 standard provides the required flexibility and scalability within each module, and it provides the glue points that allow the modules to be connected in the configuration necessary to meet the needs of any given process.

A.2   Concepts of modular manufacturing

The term 'module' is defined as a detachable section, compartment or unit with a specific purpose or function. Usually, a module is a part or a member of a larger set of similar or like items. And each member of the set has a specific function that it performs differently from the other members. However, there is a very standardized method for all the members to interface with each other.

Let us take an example of modular design from children's toys. LEGO™ bricks have a standardized means for interfacing with other LEGO bricks. LEGO bricks connect together by studs found on each piece, which fit perfectly into tubes found underneath each piece. These studs and tubes are always spaced the same distance between each other on every piece, so that when designing a new piece, the designer has the freedom to create any functionality needed for the piece as long as he designs the interface to other pieces according to the stud and tube standard. Each new member piece of the whole LEGO system of play has a unique function, allowing the designer and builder to create even more realistic and modern models. As long as they still use that same member-to-member stud and tube interface they all are inter-connectable. The idea behind this modular construction system is that you can use the same parts, each with unique functions but all with a common standardized interface, to build literally anything you can dream up.

This example is useful to understand the concept of modularity, as the concept of modularity is key to implementing a batch automation project using ISA S88 standard which also follows modular approach. For analogy, each LEGO brick can be considered as a module. It can be completely isolated from the other pieces and still perform its designed function, whether that function is just to be a square block with studs and tubes or a rocker arm assembly with stud and tube connection points. However, each piece can be connected to any other piece to form a completely different member subset module with completely different functionality than any of them alone. These pieces can then be connected to other pieces using this standard stud and tube interface to create even larger modules, or can be disassembled and reconnected differently to make a completely new subset. The key is that the basic modules are always there and never lose their independent identity, functionality and inter-connective capability as a module. A module is defined as a unique member of a larger group of items with a common interface to each other that allows any module to connect with any other module to jointly perform a greater combined function while still maintaining their individual unique identity and functionality.

As discussed in the beginning of the course, manufacturing processes are broadly classified as continuous process, batch process and discrete process. Batch and discrete processes traditionally employ the concept of modularity for process control and automation than the continuous processes. This is primarily because of the need to process material in defined quantities known as batches or lots in a defined period of time and then pass it on to the next piece of equipment. In recent times, most of the manufacturing processes have followed modular approach to take advantage of flexibility which the modularity offers. The batch and discrete process industries use the same equipment to make many different types of products just by rearranging the various process equipment modules. This is very useful particularly for the process industries where, mass production is no key of successful business.

A.3   Modular approach in batch process manufacturing

Success of a batch process design is determined by how flexible the recipes written for the given equipment, maintaining the stable rigidity of unchanging phase logic. In order to achieve recipe flexibility, it is necessary to be able to rearrange the order of activity performed by the pieces of equipment that execute each production procedure. And to have phase rigidity, it is necessary to have the phases act on pieces or groups of equipment which will always have the same function, relative to the production cycle, no matter where they occur in the recipe. In this manner, maximum flexibility can be achieved by being able to rearrange the execution of the recipe phases, without changing the phase logic or its interface to the production equipment. The purpose of modularity in batch process design is to understand and organize the activity of the equipment, into equipment entities. The understanding of how and why to modularize the batch process is absolutely critical to a good flexible recipe design. A module is viewed as a separate entity in itself. It simply executes itself in a consistent manner. Modularity indicates modules designed with a singularly common object type interface to each other. The functionality of each module remains the same, irrespective of how it is inter-connected to the other modules. Flexibility allows a simple change in the design by removing only one module and replacing it with a completely different module, utilizing the common object type interface. As discussed earlier, in batch process design we deal with modules and procedural elements that combine to make an equipment entity. These modules are identified as units, equipment modules or control modules. There are procedural elements that contain the execution logic for each module of equipment. These modules are identified as phases. The successful batch process design must appropriately define what the ideal boundaries are for modularizing the equipment, so that the corresponding execution logic remains unchanged when the recipe changes. The top-down approach leads to an optimum modular equipment design, as discussed and illustrated in the main text.

A.4 Modularity using ISA S88 standard

How often have you completed an automation project and found out only later that you could have done it a lot better if you would have known the best method. Maybe you have had to adapt the design in the middle of the project, and throw away all your earlier efforts. Thanks to the modular approach of ISA S88 standard, you will be able to adapt parts of the design, without affecting the total design. The ISA S88 standard has proved to be an affective design method, and you can take advantage of this. The modular approach of S88 provides different parts of the control of a process independent of each other. It is possible to reuse parts of your programming. Not only in one project, but also in the future, for different clients. ISA S88 modularity allows easier global replication and better return on investment. You can benefit from the replication of recipes and equipment control code. If written properly, you can duplicate equipment functionality with minimal code changes, significantly reducing the time needed to implement subsequent projects. With ISA S88, recipes are also more transportable between sets of equipment or between plants. Also keep in mind that a modular approach tends to reduce software complexity. This can lead to easier maintenance, troubleshooting and validation.

With the modular approach of ISA S88 standard, it becomes easier to make changes to processes. An adaptation to a recipe can be followed by an adaptation in the related equipment control more easily and faster if your control has been developed based on this standard. It will not be necessary to revalidate the unchanged parts of the process, because adaptations in one module usually do not effect other modules.

Another benefit that ISA S88 standard provides is the modularity of the programming code within the system. Modularity allows for easier maintenance and process modifications. Recipes and well-written equipment phase code, for instance, can be re-used in subsequent projects, effectively reducing implementation time.

Modular approach is also helpful in risk management. Let us assume that those defining batch automation systems do define early the key automation philosophies for the plant and also make full use of the ISA S88 standard structures to provide implementers with a clear catalogue of functional objects – procedures, recipes, basic controls, unit procedures, phases, etc. When we come to test the final system, we should expect to adopt this same modular approach. There are several obstacles to this. Requirements creep is probably the most significant because it is not a perfect world. Unfortunately, this is usually accompanied by divergence between the original requirements specification and the functional specification and, later on as a project change occurs, between the functional specification and the actual software functionality. Testing and validation activities are at risk because the management of the change process is often slack and the timing, content and format of test documentation is often misjudged. For validation documentation this risk is considerable since, if done badly, it can additionally cost on top of the total software implementation cost. It requires discipline in the management of change, and it requires a close working relationship between specifiers, designers, testers, validators and software engineers.

A.4.1   Process of modularization

In the process of modularization, as shown in Figure A.1, the physical model is constructed from the top-down. It documents the fundamental physical capabilities of the equipment, piping and controls. Using the basic functional capabilities of equipment modules and control modules as building blocks, the procedural model is built from the bottom-up.

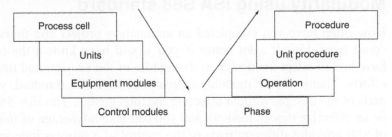

Figure A.1
Process of modularization

The process of modularization requires:

- Definition and alignment of the purpose of the module and its components. As such, almost anything that can be called a system can also be called a module. But when developing modules avoid confusing the term 'use' with 'purpose'. The term purpose defines what the equipment does.
- Use of each module must be established by defining how it interacts with other modules. Whether the module is used stand-alone, shared or is an exclusive-use module.
- Design and define module to take full advantage of equipment capability, even if not every capability is needed for every product.
- Determine if a module can be portable, duplicated or moved to another process or location. Having processes share designed, tested and proven modules is the goal of modularity and key to achieve agile manufacturing.
- Ensure modules allow capacity expansion of a process by adding and/or copying existing modules.

- Define modules to be independent in their abilities. To achieve this, the interlocks, messages and alarms necessary for the module should be an integral part of the module to operate safely and independently.
- Modules should have ability to minimize or isolate or contain the process upsets within themselves.
- Determine the physical process constraints of modules.

An example of PVC process modularization is illustrated in Figure A.2.

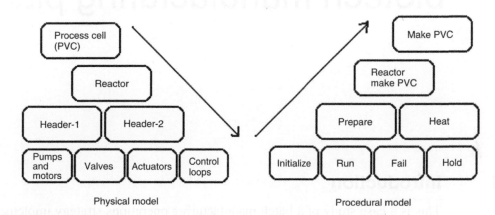

Physical model

Procedural model

Figure A.2
Example of PVC process modularization

Appendix B

Case study of batch automation in a biotech manufacturing plant

B.1   Introduction

This is a case study of a batch manufacturing operations strategy implemented at a large-scale, multi-product, multi-train biotechnology fermentation and product recovery facility. The batch management automation of this plant consists of recipe and lot number assignment, recipe scheduling and execution, material control and tracking, collection of equipment and process information, operator actions, batch product and equipment reports.

Equipments are organized in units as defined by the process model, and are automated through phases and corresponding control modules. Equipment resources are acquired and released by the recipe with equipment status – clean, dirty, in-process, etc. – which are set or changed by recipe or phase execution. Manufacturing and product quality assurance is accomplished with help of the batch control system online batch report inspections and the escalation of anomalies for subsequent resolving of incidents.

B.2   Introduction of manufacturing facility

ABC Biotech is a leading biotechnology company using human genetic information to discover, develop, manufacture and market pharmaceuticals that address significant unmet medical needs. ABC Biotech started work for a new pharmaceutical manufacturing campus in early 1990s with the development of manufacturing requirements and the search for a suitable plant site. The selected site-A is about 100 km from ABC Biotech head office. The manufacturing facility is designed for large-scale, cell culture fermentation and recovery production. It has the capability and flexibility to produce approved products, as well as products which are in the product development pipeline. A product for breast cancer approved by the FDA is manufactured at this facility. Automation was a necessity for better control of the large amount of sophisticated equipment and complex processing requirements. The engineering team had faced challenges to develop and communicate a technically feasible design concept to the various stakeholders to obtain their acceptance. Conceptual engineering was started and the manufacturing facility was mechanically complete in 2 years. Startup and qualification activities took about 6–9 month period.

Presently, the manufacturing facility consists of five buildings. The manufacturing building houses processing equipment for large-scale, multi-product, multi-train cell culture fermentation, recovery and purification processing equipment. Utility systems and other supporting buildings are located at strategic locations to service the manufacturing process. All buildings are situated along an enclosed spine, which connects the building utilities, materials and people flows.

The manufacturing building has three processing levels and a penthouse for HVAC utilities and other mechanical building infrastructure equipment. The process building contains several intermediate product cells for preparation of media, buffers and other components. Several cell culture fermentation operations are located in one part of the building, while the product is recovered and purified, in a single train, in the other part of the building. All processing areas are kept very clean and sanitary, most are controlled to class 100 000, some to class 10 000 standards. The manufacturing is organized in processing areas, which are separated by gray spaces. These are utility spaces which contain piping, pumps, instrumentation and other equipments not required in operating areas.

The facility is complex; all activities are computer-controlled, including manual activities. Modular design techniques, as described in the ISA S88 standard, were implemented to obtain a high degree of flexibility in equipment and process operations. The process building contains over hundred processing vessels located in several operational areas or cells such as, fermentation, media make up, buffer make up and buffer hold, recovery and purification.

The DCS contains over 6000 I/O points, distributed over 50 controllers. Process operators interface with the process using over 200 different graphic display pages on 20 HMIs which are located at strategic processing locations, distributed throughout the plant. There is no central control room, since the batch operations require many manual or computer-instructed operations which have to be executed at the equipment on the manufacturing floor. In addition to the DCS controllers, the DCS interfaces to a dozen PLCs as part of vendor-supplied equipment skids for monitoring and control. The DCS is integrated with the enterprise managerial and control systems.

Building HVAC and control utilities, such as boilers, chillers, cooling towers, etc. is independent from the DCS, controlled by a separate building automation system (BAS). The BAS and the DCS are communicating through a gateway to share process utility and alarm information. Over 3000 test procedures have been executed to qualify and validate the DCS automation system.

B.3 The manufacturing process

The manufacturing process is designed for an output of about five batches per week. It is necessary to execute over 200 different recipes to prepare equipment and intermediate products to produce one batch of product. The process starts in the cell culture lab where biologically engineered cells are taken from a cell bank and grown in laboratory scale fermentation process until enough cells are produced to seed the first industrial fermentor. Bioengineered cells are grown to high cell densities in successive size fermentors until large quantities are obtained. Thereafter the cells are 'turned on' to produce the required protein or product. The harvest operation separates the product from the cells, whereafter the cells are killed and discarded. The product undergoes several recovery and purification operations. This is accomplished in large-scale filtration and chromatography operations. A large part of the manufacturing activities is the formulation and preparation of growth media's nutrients and the necessary buffers to support the production process. Process flow is illustrated in Figure B.1.

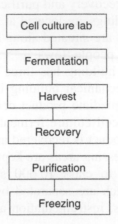

Figure B.1
Product manufacturing process

B.3.1   Complex batch

In this facility the product procedure is executed and controlled by a single recipe. The recipe procedure consists of a dozen major process operations, which acquire and release units successively, as required to obtain a finished product.

Complex batch is defined as:

- Most recipes require the interaction of multiple units simultaneously.
- About 200 recipes are required to prepare equipment, and to produce the necessary intermediate products to produce one batch of product.
- Recipes need to be scheduled as part of product production procedure. The scheduling should be such that equipment with the desired status is available, just in time, for the product recipe for continuing process operations.

B.3.2   CIP and SIP

Preparation of equipment process is illustrated in Figure B.2. Processing vessels are automatically cleaned and sterilized in place by CIP and SIP systems. It is a requirement to automate the CIP and SIP operations, since it would be uneconomical and would require too many resources and time to clean these systems manually. Clean steam is supplied in a header to the vessels. Cleaning solutions are supplied from several CIP skids through a CIP distribution network. CIP is fully automated in order to meet the cleaning demands and schedules. It is imperative that product processing be automated to the same degree as the CIP, since most valves in the process piping are automated as required by the CIP and SIP operations. Product transfer piping is cleaned and sterilized before product transfer can commence. To maintain flexibility of equipment resources, transfer panels are employed to route transfers, under sterile conditions, from any upstream vessel to any downstream vessel.

Figure B.2
Preparation of equipment

B.4   The automation project objectives and requirements

Following were the automation objectives and requirements for the process:

- Build a flexible multi-product cell culture batch manufacturing facility
- Automation of repeatable process operations and reproducible product
- Reduce cycle time. Collect required process data to enable efficient and reliable release of product
- Reduce risks.

In the past, process operations used paper tickets, which contained the formula and operating instructions, filled out and signed manually, and became the paper batch record. The vision was to replace the tickets with automated instructions to both automated equipment and to technicians for manual activities. During the vision development, it was found that automation and process control scope far extends beyond the automation of equipment. Control of equipment and product through automated procedure is the key part of the plant's manufacturing execution strategy. Additional interaction with other automation systems, such as MES, LIMS, ERP and documentation management became a requirement.

B.5   The batch system requirements

Following batch control system requirements were identified:

- Recipe-controlled procedures to obtain product consistency and reproducibility including fast changeover between products.
- Modular equipment units to create flexibility so that additional equipments can be added without significant software changes to the existing system.
- Automate the operating technician as well as equipment to obtain consistency of operations.
- Electronic signatures conforming to FDA regulations; operating technician and/or QA signs after each significant process activity to create the batch record on a parallel path to reduce cycle time.
- Recording and sign off of all significant activities.

- Tracking of material ID and lot numbers to create the lot trace.
- Reporting of anomalies; manufacturing quality assurance (MQA) reports are generated, to report and resolve anomalies as they occur, to reduce the product approval cycle.
- Automatic batch report generation at the completion of the recipe.
- In parallel with the manufacturing of product, product lot information is created for the disposition and release of the lot.

B.5.1   Challenges

During the project implementation, the S88 standard was just approved. However it was a great help to purvey the definition of terminology and the design of the process model, to process engineers and management, Sections 5.2.3 and 6.13 of the S88 standard indicated that structuring of equipment entities and process engineering of recipes is a complex and iterative engineering activity. This activity went through many brainstorming iterations, and took up much more time than anticipated, before the strategy of recipes operations was defined. With the strategy mapped out, the design and definition of equipment entities, such as unit boundaries, dedicated and shared equipment modules, and connections is much easier to accomplish.

To control product through the many procedural activities in numerous equipment units, it was required to switch from equipment-centric control to product-centric control. Under product control, the recipe stays with the product, and can acquire or release several units as required at the same time in one recipe. It is sometimes difficult to realize that it is the product that needs to be controlled and produced, not necessarily the equipment.

Though the DCS had limited batch functionality available, and through vendor partnering, many additional functions, essential for batch execution where developed. Some of the additional developed functions are:

- *Unit status*:   This is a label on the unit which identifies the disposition of the unit with information such as unit number, clean, dirty, batched, acquired by lot number, the batch number, time of operation, run, hold, etc. Units now can be acquired by status.
- Electronic signature with conformance to FDA regulations, to be used by anywhere in the system.
- *Material use*:   Information entry with use of barcode and e-signature for lot trace.
- Electronic signature and operator comment after a control module is placed in automatic control.
- Maintain process filter status and tracking.
- Expiration dating and automatic degrading of equipment status, necessary for intermediate products, CIP and SIP.
- Integrated information sharing with MES system and batch reporting.

From the initial concept design, various individuals from manufacturing and other departments were included as members of the team, which helped tremendously in the communication, justification and acceptance of these new concepts. This resulted in enthusiasm during the development work, and achieved an overwhelming buy in and ownership by the various departments and stakeholders.

B.5.2 Recipes

Many process operations must be executed concurrently with the product recipe to produce product including preparing equipment such as cleaning after the previous recipe has been completed sterilizing and the formulating of intermediate products for use in the production process. Three types of recipes are defined to accommodate the plant requirements and work flow:

1. *Product recipes*: These are product-specific and change the product status. These recipes stay with the product throughout the facility unit until the product is finished.
2. *Intermediate recipes*: These are specific to an intermediate product, such as media, buffers, etc. These recipes stay with the intermediate product, until the product is finished.
3. *Equipment recipes*: These are equipment-specific, and are used to prepare equipment for subsequent process operations, such as CIP, SIP, pre- and post-use of product procedures, such as prime, equilibrate, etc. These recipes change the status of equipment, they may use intermediate materials, but they do not make a product.

The acquisition of equipment units necessary for the various recipe procedures and operations is illustrated in the Figure B.3. The 'Make Product' recipe acquires and releases equipment as it moves through the process operations.

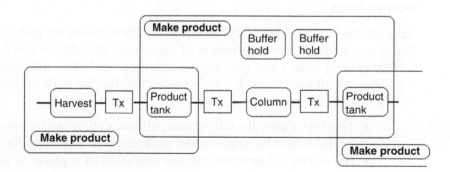

Figure B.3
Recipes

B.5.3 Tracing equipment and materials

The status and contents of equipment is tracked by means of an electronic unit label, which identifies the lot number that produced the material. Raw materials and intermediates used in the process are read by barcode, verified for accuracy and are tracked to produce a required lot trace.

Data and information that is being generated is related to recipe lot numbers and historized. The next recipe, that acquires the equipment and/or the material, also acquires the lot number that produced the unit status or the material to track genealogy.

B.5.4 Recipes acquire and release unit

Execution of recipes or phases changes the status of equipment or product. Equipment resources have a state (auto, semi-auto and manual) and a status that is used by the recipe to acquire equipment or product for subsequent operations.

Equipment status is defined as follows:

- Available and dirty
- Clean – clean expired
- Sterile – sterile expired
- Media or buffer batched
- Batched expired
- Other intermediate material status and qualifiers are defined as necessary
- In use – processing
- Out of service – maintenance, etc.

Only equipment resources that are 'released' and 'available', with a certain status, can be acquired by a recipe for process operations. A CIP recipe may acquire equipment with 'Dirty' or 'Clean' status. A product recipe may require that only equipment with 'Sterile' or 'Batched' (with an intermediate product) status is acquired. Equipment that degrades to an expired state such as 'Sterile Expired' cannot be acquired by a recipe that requires 'Sterile' equipment. 'Sterile Expired' equipment must be resterilized before it can be acquired and used in an product recipe. Additionally, the status of connections, i.e., jumper position of transfer panels, and transfer paths with one or more piping segments is individually tracked.

B.5.5   The process model defines equipment units and other modules boundaries

Acquisitable of units, equipment modules and connections are part of the process model. The process model is the definition of equipment boundaries and functionality. In this facility several units must typically be acquired, including all piping connections and segments, which need to work together to accomplish the required process operation. For example:

- To clean a unit, the CIP system, all connections and corresponding equipment modules must be acquired and work together to perform the cleaning operation.
- Harvesting a fermentor requires execution of a phase in the fermentor, in the harvest unit and in appropriate buffer units and other hold and transfer vessels or units.
- Chromatography operations require a chromatography column unit, several buffer tanks and product vessels, etc.

The process model gains flexibility through definition of units and equipment modules granularity, such that they can be acquired by the various product or equipment recipes without interacting or interfering with each other to perform their respected process operations. For example,

- A fermentor until that is 'charged' with media, through an intermediate product recipe, is released as 'Available and Media Batched'. The fermentor will be acquired by a product recipe to be inoculated. However, the media supply line, the 'connection' between the make up tank and the fermentor, is made up of separate segments, including the transfer panel. Segments may have their own equipment modules. The make up vessel and the piping connection at the fermentor is released as 'Available and Dirty' and will be acquired and cleaned by a CIP recipe, without interfering with ongoing operations in the fermentor unit.

- A buffer hold tank must be acquired, as 'Clean', by a intermediate 'Make Buffer' recipe to be filled with buffer. It will be released with the status 'Charged', and subsequently will be acquired by the column preparation recipe. When the column preparation is completed, it will be released again with status 'Prepped', and will subsequently be acquired by the product recipe for product operation.

B.5.6   Phases change, and control modules maintain status

Recipes acquire units and instruct phases. Phases receive control parameter instructions from the recipe and instruct control modules. Phases change the status of equipment or product by directing control modules (example: an environmental phase in a fermentor unit sets the parameters of several control modules). Control modules maintain status of equipment or the contained product. When a unit is released, some control modules will be turned off, others remain active and maintain status of the equipment. For example, temperature and pressure control of an intermediate material what is released and available to be acquired by the product recipe. Note, that the unit now is released, control modules are active and maintain status, and it retains the lot number that created the material. Additional control modules were developed to maintain the status of filter elements, and other mechanical components or consumables used by the equipment in process operations. Item and track numbers are tracked for lot trace reporting.

B.5.7   Role and responsibility of the operating technician

In automatic mode the DCS is in charge and execute recipes, phases and instructions to operating technicians. The technician is alerted and instructed to perform an activity, and when completed, enter the information along with electronic signature. Direct the control system actions in semi-automatic or manual mode. The technician may comment on actions taken. It is required to comment on why was the system put in manual and electronic sign off, when a control module or phase is placed back in automatic mode. In a quality assurance function, verification and electronic sign off of automated control actions or resolving of exceptions may be required.

B.5.8   Batch history reporting

This is a good manufacturing practice operation and all significant activities are recorded, evaluated and approved for accuracy and completeness. Automated and manual activities are under recipe control by the DCS. Automated phases are automatically confirmed. Manual phases, through instructions to operators, require electronic signatures by the technician to confirm the action.

Recipe instructions and electronic recorded information is located in a single relational data base, which replaces and eliminates the use of tickets in manufacturing. This gives the manufacturing quality assurance group a window in the process for, quickly and accurately, establishing the disposition of the batch. Anomalies, such as critical alarms, are identified early and are resolved in a timely manner which increase product lot release and reduce cycle time.

B.6   The system architecture

At the center of the DCS is a redundant pair of DEC/Compaq 4100 Alpha servers and a disk array configured in a tru-cluster. The system architecture is illustrated in Figure B.4. The machines are cross-strapped with dual 100 mb Ethernet connections to communicate to the HMIs, controllers and between the servers themselves. Only one server is active, and is scanning and updating the controllers. A mirroring program maintains the update between the real-time data bases on the two servers. The batch manager and historian programs are running on the active server only. They are dormant on the standby server and will only be activated during a switch over when the active server goes down and the standby server becomes the controlling server.

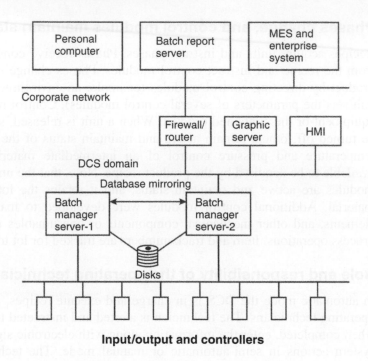

Figure B.4
System architecture

There are 20 HMIs located on the manufacturing floor which get their graphics from eight distributed graphic servers. All control application code such as control modules, phases, unit-control, etc. is located in unit controllers, which are with their I/O modules located in centrally located I/O rooms. Control language is 4-mation and conforms with IEC 61131-3 standard. Control recipes are executed on the Alpha servers which contain the process model and the process supervisor. Batch and trend historian data is off loaded and historized on separate report servers to minimize the load of the DCS control system.

The DCS is considered a closed system, no users on the corporate intranet are allowed in the DCS domain. A router is set up as a firewall between the DCS domain and the corporate intranet for certain approved communications. A batch report server is located on the corporate intranet, and is utilized to pull, under secure conditions, certain batch reports off the batch historian and route this to the requesting user. A similar server is set up to make historical trends available for others in the company to use. A MES/DCS interface is used to send information of materials, used in the process, to a MES system for material genealogy tracing and inventory adjustment.

B.7   The plant startup and validation

A commissioning plan was developed detailing the various startup tasks and the responsible party for execution. Control qualification (CQ) of hardware and software started at the initial development activities with the testing of the control module library.

Control modules and phases were integrated in units, which were tested and duplicated as clones. Testing of hardware, wiring, fusing and calibration was formally performed by the installation contractors. Over 3000 tests were performed to complete all the testing of the DCS. A change control system was implemented, which consisted of impact assessment, implementation testing and peer review, documentation update, sign off and close out of the requested change during startup. This enabled us to stay in control of the software, while we had at some stages forty engineers working on the system.

Due to the large system size and the enormous amount of data that needed to be scanned and updated, we were challenged with numerous problems with system communication and other system softwares. Working with the control vendor, in a close partnership, we worked systematically through all issues to increase system performance and robustness. Several startup teams were formally formed with corresponding team leaders. Separate engineering and CQ teams were dedicated for support of startup team activities.

B.8 Conclusion

The project implemented in this case study was another step in the evolution of automated batch process control technology. This was a large and very complicated batch process. And implementation of ISA S88 standard Part 1 concepts added control functionality, repeatable process operations and product consistency. Flexibility is obtained by detailed process model design. Additional system functionality is added to track equipment and product status to obtain a higher degree of production success rate. Cycle time and release of product is improved by efficient process data presentation, review and anomaly resolution. The end result of using advanced technology is a successful, flexible and agile manufacturing facility fully automated and controlled by execution of recipe instructions to automated equipment and to operator technicians for manual interactions.

Solution 1

[A] Fill in the blanks to complete the following statements:

1. Based on the output of the process, industrial processes are classified as batch process, continuous process or <u>discrete</u> process.
2. Economy of scale, as a key to the success in business in chemical/process industries led to focus on designing and developing <u>continuous</u> processes.
3. Manufacturing of fine and specialty chemicals with the increased emphasis on high quality and customer requirements led to focus on <u>batch</u> processes.
4. Based on structure, batch processes are classified as <u>single-path</u>, <u>multiple-path</u> and <u>network</u>.
5. A multi-grade batch plant produces products that are similar but not <u>identical</u>.

[B] Identify the following processes as either batch process, continuous process or discrete process:

1. Assembly of watches [Discrete]
2. Soap manufacturing [Batch]
3. Beer brewing[Batch]
4. Cement manufacturing [Continuous]
5. Manufacture of anti-tuberculosis drug[Batch]
6. Production of cars [Discrete]
7. Manufacturing toothpaste[Batch]
8. Electricity generation[Continuous]
9. Paper manufacturing [Continuous]
10. Assembly of television sets   [Discrete]

[C] State these statements as either TRUE or FALSE:

1. Batch manufacturing plants are comparatively more robust than a continuous plant.   [True]
2. Batch process manufacturing facility is easy to scale up depending on market demand and requirements.   [True]
3. In a continuous process, each of the equipment operates in a single steady state and performs specific processing function.   [True]
4. Batch process plants can be made flexible and are suitable for producing special products with few equipment.   [True]

5. A multi-product, network-structure batch process is the most complex batch
   process.   [True]
6. The ISA S88 Part 1 standard is a compliance standard.   [False]

[D] Arrange the following elements of the physical model as described in the ISA S88
     standard in hierarchical order (top to bottom):

Area, Enterprise, Control module, Equipment module, Process cell, Site, and Unit.

Solution:

Enterprise
Site
Area
Unit
Process cell
Unit
Equipment module
Control module.

Solution 2

[A] The physical model described in the ISA S88 Part 1 standard has the following elements:

Area, Control module, Enterprise, Equipment module, Process cell, Site, Unit.

1. List elements of the physical model covered by scope of the ISA S88 Part 1 standards:

   Process cell
   Unit
   Equipment module
   Control module.

2. Element of physical model that is a domain for a batch control system is:

   Process cell.

3. Physical model elements which may contain element(s) of same levels:

   Equipment module
   Control module.

4. Physical model element for carrying out major processing activity:

   Unit.

5. Physical model element that performs basic control:

   Control module.

6. Physical model element for carrying out minor processing activity:

   Equipment module.

[B] Tick from the following, those you will identify as a unit:

   1. Reactor   [Unit]
   2. Pump   [X]
   3. Mixing tank   [Unit]

4. Ingredient storage tank   [X]
5. A refrigerator   [X]
6. A washing machine   [Unit]
7. A kitchen blender   [Unit]

[C] State the following statements as either TRUE or FALSE:

1. Physical model is collapsible.   [True]
2. Batching can occur without units.   [False]
3. Control modules are defined in the batch management system.   [False]
4. A process cell is usually independent of product.   [True]
5. A process cell must contain at least one unit.   [True]
6. A control module does not have to be a part of an equipment module to be a part of unit.   [True]
7. Unit can contain more than one batch at a time.   [False]
8. Equipment modules execute portions of a recipe, but control modules do not. [True]

[D] Identify equipment modules (EM) and control modules (CM) from the following:

1. Weigh scale   [CM]
2. An inkjet printer for labeling [EM]
3. Hoist   [EM]
4. A damper actuator   [CM]
5. A sampler   [CM]

[E] Figure E2.1, illustrates an example of a physical model of an ABC Enterprise.

Answer the following questions based on the physical model of the ABC Enterprise:

1. Number of process cells in the ABC enterprise?

   4.

2. Total number of units in the ABC enterprise?

   5.

3. Process cell(s) for which no unit(s), equipment module(s) and control module(s) are illustrated in the physical model?

   Sterilization, Utilities.

4. Unit(s) for which no equipment module is illustrated?

   Accumulator.

5. Unit(s) for which no control modules are illustrated?

   Accumulator, Feed tank.

Solution 3

[A] List down the elements of process model in hierarchical order:

Process
Process stage
Process operation
Process action.

[B] State whether the following statements are TRUE or FALSE:

1. A process stage, usually operates independently from other process stages. [True]
2. A process operation describes a minor processing activity. [False]
3. In a process model, the procedure for making a product does consider the actual equipment for performing the various process steps. [False]
4. A process operation results in chemical and/or physical change in the properties of the material being processed. [True]

[C] Figure E3.1 illustrates an example of process model for raw material slurry batch preparation process for manufacturing fiber–cement sheets.

Answer the following questions for the illustrated process model in Figure E3.1:

1. How many process stages are there in the process model?

   3.

2. Identify the process operations in each process stage?

   Fiber slurry preparation

   (i) Mix/stir fiber to open-up
   (ii) Mix pulp with fiber

   Cement slurry preparation

   (i) Mix cement

   Fiber–cement slurry preparation

   (i) Mix fiber slurry and cement slurry.

3. Identify the process action(s) which is common in all the process stages in Figure E3.1.

   Add water.

Solution 4

[A] Fill in the blanks to complete the following statements:

1. Procedural control is a link between <u>process</u> model and <u>physical</u> model.
2. Two types of procedural elements are <u>recipe</u> elements and <u>equipment</u> elements.
3. <u>Procedure</u> defines the strategy for accomplishing a major processing action, like making a batch.
4. <u>Operation</u> is a sequence of phases that defines a major processing sequence that changes the state of the material being processed.
5. <u>Basic</u> control includes exception handling, interlocking, sequential control, regulatory control and alarm annunciation.

[B] State whether the following statesments are TRUE or FALSE:

1. Procedural control involves control recipe procedure and equipment control. [True]
2. Procedural control directs equipment-oriented actions in an ordered sequence to carry out process-oriented tasks.   [True]
3. A recipe procedural element is independent of the equipment on which it is executed.   [True]
4. More than one unit procedures can be active in a unit at a given point of time. [False]
5. An operation usually involves a chemical or physical change.   [True]

[C] Identify the blank elements (boxes) in the following diagram to complete the sequence of activities:

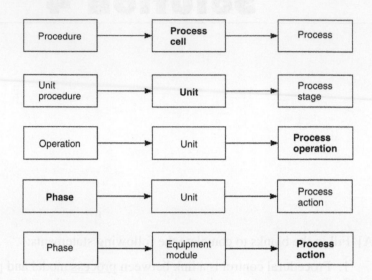

Solution 5

[A] List down the missing components of a recipe in the following diagram:

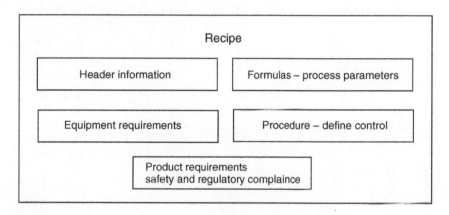

[B] Match the following:

1. General recipe   A. Accounts for specific information like equipment arrangements necessary to make the product

2. Site recipe   B. Contains complete scheduling, operational and equipment information

3. Master recipe   C. Describes the requirements to manufacture a product at various sites

4. Control recipe   D. Language, units of measurements and raw materials are adjusted

[C] State the following statements as either TRUE or FALSE:

1. A recipe should be independent from the equipment on which it is executed. [True]
2. General recipe is created with specific knowledge of the process cell equipment that will be used to manufacture the product. [False]
3. Master recipe is targeted to a specific process cell. [True]
4. A master recipe can be created as a standalone entity without a general recipe or a site recipe. [True]
5. Usually, a recipe procedural element maps to an equipment procedural element at phase level. [True]

[D] Example of a typical recipe to manufacture toothpaste is shown in Figure E5.1.

Answer the following questions based on the recipe.

1. How many unit procedures are involved in the manufacturing process?

    3.

2. Does Figure E5.1 illustrate all the operations involved in the manufacturing of toothpaste?

    No.

3. Identify the phases shown in Figure E5.1.

    Add water, add NaF and add fillers.

Index

Printed and bound by CPI Group (UK) Ltd, Croydon, CR0 4YY

03/10/2024

01040338-0017